STARS OF THE
SOUTHERN SKIES

Dedicated to my mentor, the late Professor Arthur Bleksley

STARS OF THE SOUTHERN SKIES

An Astronomy Field Guide

Mary FitzGerald

WITS UNIVERSITY PRESS

WITS UNIVERSITY PRESS

Witwatersrand University Press
1 Jan Smuts Avenue
2001 Johannesburg
South Africa

ISBN 1 86814 410-0

Cover picture: Crab Nebula
All star charts and diagrams by Frank Flowers
Design and layout by Mad Cow Studio, a division of STE Publishers
Printed and bound by Paarl Print

Contents

Introduction   2

Acknowledgements   3

Light and Telescopes   5
  Light   6
  Telescopes   8
    Refracting   9
    Reflecting   9
    Catadioptric   10
    Oculars (eyepieces)   10
    Viewfinder   11
    Telescope mounts   12
    Computer-controlled telescopes   12

Bright Stars and Constellations   13
  Star magnitudes   15
  Measuring distance   16
  Galaxies   17
    The Milky Way   20
  Novae   24
  Double and multiple stars   25
  Star maps   27
    The evening sky of summer: looking north 28
      Orion (The Mighty Hunter)   29
      Canis Major (The Great Dog) 31
      Canis Minor (The Little Dog) 31
      Taurus (The Bull)   31
      The Pleiades (M45)   32
      Pisces (The Fishes)   33
      Auriga (The Charioteer)   33
    The evening sky of summer: looking south 34
      Tucana (The Toucan)   35
      Hydra (The Water Snake/Sea Serpent) 35
    The evening sky of autumn: looking north 36
      Gemini (The Twins)   37
      Cancer (The Crab)   38
      Leo (The Lion)   38
      Coma Berenices (Queen Berenice's Hair) 39

The evening sky of autumn: looking south 40
    Corvus (The Crow) 41
    False Cross 41
The evening sky of winter: looking north 42
    Virgo (The Maid/Virgin) 43
    Libra (The Scales) 43
The evening sky of winter: looking south 44
    Centaurus (The Centaur) 45
    Scorpius (The Scorpion) 48
    Sagittarius (The Archer) 49
The evening sky of spring: looking north 50
    Pegasus (The Winged Horse) 51
    Delphinus (The Dolphin) 51
    Cygnus (The Swan) 53
The evening sky of spring: looking south 52
    Grus (The Crane Bird) 53
    Piscis Australis (The Southern Fish) 53

Comets and Meteors – *Tim Cooper* 55
  Comets 56
    Where do comets come from? 56
    The anatomy of a comet 58
    Comet appearances 60
    How to observe a comet 61
  Meteors 62
    How to observe meteors 65

The Sun and the Moon 67
  The Sun 68
  The Moon 69
    Phases of the Moon 74
  Eclipses 75
    Lunar eclipses 76
      Types of lunar eclipses 77
      Magnitude and duration of a lunar eclipse 77
      Circumstances of a lunar eclipse 78
      Darkness and colour of a lunar eclipse 78
    Solar eclipses 79
      Types, duration and circumstances of a solar eclipse 80
      Observing a solar eclipse 80
      Forthcoming solar eclipses 82

The Planets   83
    Mercury   88
    Venus   90
    Earth   92
    Mars   94
    Jupiter   96
    Saturn   100
    Uranus   104
    Neptune   106
    Pluto   108
    Sedna   110
    Proplyds   111
    The Asteroids   111

Further Reading   114

Appendix I
    Constellation Names   115

Appendix II
    The Greek Alphabet   118

Appendix III
    Observatories, Societies and Places of Interest   119

Introduction

Introduction

Since time immemorial, humankind has watched and wondered at the seasonal procession of the stars across the heavens, and the rhythmic wanderings of the planets among the constellations of the zodiac. The night sky is accessible to all, and it is here that an enjoyable start can be made to comprehending the beauty, as well as the history, of the oldest science – in fact, astronomy is one of the few sciences that allow for active participation by non-professionals.

In many urban areas the glow of civilisation has caused a dimming of our skies and a decrease in our enthusiasm for looking upwards and enjoying the majesty of this natural phenomenon. For the city viewer, however, there is still a wealth of discovery to be undertaken amongst the brightest constellations – and having once looked up, a great many people want to lend wings to their exploration by using a telescope or binoculars.

It has been said that those who have never looked through a telescope have no adequate idea of the wonders of the heavens. It is also true that some readers who study the various specimens of fine celestial photography made with the Hubble Telescope and large observatory telescopes are apt to think they can have the same view through almost any telescope. Hopefully the disappointment will be brief!

This book is intended primarily for amateur observers of the southern sky. It contains invaluable background information on equipment and astronomical principles as well as star maps especially prepared for beginners and city viewers. Using this book beginner astronomers will be able to identify the constellations of the southern skies, and from there it is only a step to pinpointing individual stars many light years away. The star maps selected for this book contain only the brighter stars that are visible from urban areas. The different sized dots represent the various magnitudes or brightness of stars. The planets, Sun and Moon are also covered, as are asteroids, comets and meteors, with invaluable pointers to where and when to look.

The book is designed both to be read at leisure and to aid observation in the field. May it reveal to you not merely a new world but an unsuspected universe!

Acknowledgements

It is a great pleasure to acknowledge with gratitude my debt to Tim Cooper for writing the section on Comets and Meteors. Tim Cooper is the Director of the Comet and Meteor Section of the Astronomical Society of Southern Africa (ASSA) and a Council Member of ASSA. He is a Voting Member of the International Meteor Organisation, a member of the International Astronomical Union and the South African correspondent for the International Comet Quarterly. Above all, he is a personal friend, and we have spent many a viewing evening on a bushveld farm counting meteors.

I would like to express my deepest thanks to Frank Flowers for his superlative drawings of the star charts, diagrams, and so on. Without his colourful charts and diagrams this book would be utterly dull. Frank Flowers is a former lecturer at the Planetarium of the University of the Witwatersrand. My sincere thanks to Dr. François du Toit, an enthusiastic amateur and astro-photographer for his impressive astronomical photographs.

To Frances Perryer, who edited the manuscript and made several valuable suggestions for improvement, and to Pat Tucker, who proofread the book, heartfelt gratitude; and a deep bow to Margarethe Mostert, Commissioning Editor, Wits University Press, without whose trust and patience this new edition would not have been possible.

Light and Telescopes

Light

Galileo Galilei (1564-1642)

Sir Isaac Newton (1642-1727)

Traditionally telescopes detect visible light. Light from distant objects is brought by their lenses or mirrors to a focus at which the image is viewed or photographed. Galileo and Newton made important contributions to our modern understanding of light as well as to our theories of gravity and mechanics. Galileo made one of the first attempts to measure the speed of light, but the first reliable measurement was made in 1675 by Olaus Roemer, a Danish astronomer who studied the motions of the moons around Jupiter. More accurate measurements were made in the mid-1800s. From all these experiments we now know that the speed of light in a vacuum is about 300 000 km/sec. (It travels at a slightly slower speed through a dense substance such as glass, for example, but on exiting the substance it resumes its original speed.)

Whatever the nature of light, it does seem to travel somehow from a source to our eyes, and it travels very swiftly. We see a distant event before we hear the accompanying sound.

A major breakthrough in understanding light came from a simple experiment performed by Sir Isaac Newton. In the late 1600s Newton found that a beam of light passing through a glass prism is spread out into the colours of the rainbow. This rainbow, called a spectrum, proved to Newton that white light is actually a mixture of all colours.

Until recently all information about the universe gathered by astronomers was based on ordinary visible light. But with the discovery of non-visible electromagnetic radiation, scientists discovered objects emitting radio waves, X-rays, and infrared and ultraviolet radiation. Non-visible wavelengths have greatly enhanced our understanding of the cosmos.

Some interesting phenomena in the night sky can be observed without the aid of a telescope. These are not, truly speaking, astronomical phenomena, but are related to the passage of light from objects in the sky through the Earth's atmosphere. The twinkling of the stars is caused by disturbances in our atmosphere. When starlight passes through the atmosphere it is diverted slightly from its original path, hence there appears to be a small displacement of the star in the sky. This leads to the rapid and irregular changes of position and brightness we know as twinkling. An air column of great density may cause a convergence or divergence of the rays coming to the eye. This refractive effect is more marked for the blue than for the red rays. In the case of very bright stars, like Sirius, this separation of the light of different colours may be visible as flashes of different coloured lights in time with the twinkling.

A swallow caught in the light of a sun halo or sun dog

F. du Toit

Stellar objects are not the only lights seen in the heavens, but the rest – haloes, mock-suns, rainbows and so on – are much lower-level phenomena and really belong to meteorology, the science of weather, rather than to astronomy. A halo, for instance, is produced by a very thin cloud known as cirrostratus, made up of a sheet of ice crystals at high altitude. The ice crystals break up the light from the Sun (or Moon) into its component colours and a halo results. Very often the cloud itself is so tenuous that it cannot be seen directly, and it betrays its presence only by the halo it produces. Farmers say that a halo means approaching bad weather and they are often right. In fact the halo itself has nothing to do with the weather; but the halo-producing cloud is often the forerunner of rain.

The blue of the sky is caused by the scattering effect of air molecules and dust. Blue light is easily scattered; red light, which has a long wavelength, is much more penetrating. The light of the Sun is made up of a blend of all the colours of the spectrum, and the shorter waves become scattered, so that the bluish light is spread all over the sky.

Scientists also use light as a tape measure for the galaxies and other deep sky objects. The distances and sizes in the universe are so great that our normal units of measurement – metres and kilometres – are quite inadequate to express them. Thus, a new unit of measure was devised – the light year, based on the distance which a light-wave, speeding at approximately 300 000 km/sec (in a vacuum), would travel in a year. It has proved rather clumsy, as these days, with modern discoveries, astronomers may speak in hundreds, thousands and even millions of light years. Nevertheless, it is still in use.

Other units of measure used in astronomy are:

AU – The average distance of the Earth from the Sun is one of the standard units of measure in astronomy. It is called the Astronomical Unit (AU), and according to the latest determination measures 149 597 870 km. It is normally rounded off to 150 million km.

Parsec – The parsec equals 3,3 light years or 206 000 AU, or 30 857 million million kilometres.

Telescopes

Although it is possible to learn a great deal of astronomy without a telescope, the possession of even a very small instrument is a great advantage. The telescope is the only means by which we can even begin to grasp the vast significance of the universe. Even binoculars are somewhat better than the eye. But do avoid very small telescopes (50 mm objective or smaller): they are a waste of money from an astronomical point of view. Rather purchase a good pair of binoculars until sufficient funds are available for a better telescope.

The telescope functions in two main stages, first gathering a large amount of light from a distant object (far more than the human eye alone can) and focusing it to form an image, and then magnifying the image so that the object can be seen better. The telescope's performance depends primarily on the diameter of the main lens or curved mirror. The diameter is usually referred to as the aperture. The larger the aperture, the more light is gathered, and the better the image will be. Aperture is usually measured in inches. Another important determinant of a telescope's performance is the focal length, which is simply the distance from the lens or main mirror's surface to the point where the light comes to a focus. The two terms (aperture and focal length) are sometimes combined in the f-ratio, which is just the focal length divided by the aperture. Therefore, if a telescope has a 6-inch mirror and the focal length is 48 inches, its f-ratio (or f) is f/8.

The properties of the image produced by a telescope, whether by means of mirror or lenses, are similar. The real image produced is inverted – that is, top and bottom are reversed, as are left and right. An additional lens, called an erecting prism, can be used to invert the image a second time so that objects will appear as they do when viewed with the unaided eye.

There are basically three types of telescopes refracting, reflecting and catadioptric.

Frank Flowers

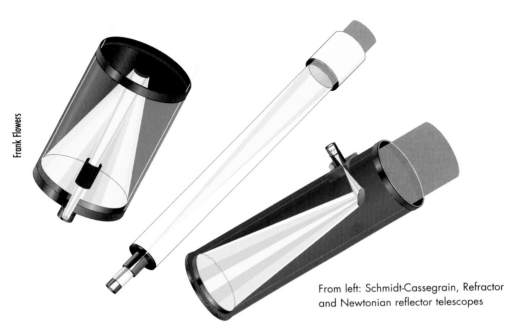

From left: Schmidt-Cassegrain, Refractor
and Newtonian reflector telescopes

Refracting

The refractor's principal components are the lens (objective) mounted at one end of
the tube, and, at the other end, a small lens combination forming the eyepiece through
which the observer looks. Two optical dimensions of the lens are of special interest: its
diameter and its focal length. The diameter of the lens determines, among other
things, the capacity of the telescope as a light collector. In most well made refracting
telescopes almost all the light caught by the lens reaches the eye. In the refractor the
large lens collects the light and forms an image near the end of the telescope. The
image is then magnified by an eyepiece (ocular). Different magnifications can be
obtained by selecting eyepieces of differing focal lengths.

Because different colours of light are bent differently (known as chromatic
aberration), a refracting telescope's two main lenses must be made of crown and flint
glass to correct this unequal bending. In addition, all the glass surfaces must be coated
with a thin layer of magnesium fluoride or zirconium, which enhances contrast and
reduces unwanted reflections.

Reflecting

The reflector built by Sir Isaac Newton in 1672, called the Newtonian, is probably the
most popular kind of reflecting telescope. It uses a specially curved aluminium-coated
glass mirror to gather and focus the light, and a flat secondary mirror to direct the
light out of the side of the telescope tube and into the eyepiece. Chromatic reflection

does not occur with the reflector, as light is not passed through the mirror, but reflected off the aluminised surface. The Newtonian is an inexpensive, fairly simple design, which can easily be constructed by the amateur astronomer. However, the erect image prism cannot be used in conjunction with this telescope.

Catadioptric

Another form of the reflecting telescope is the Cassegrain. It has the advantage of placing the focal point at a convenient and accessible location. A hole is drilled directly through the centre of the primary mirror. A convex secondary mirror placed in front of the original focal point is used to reflect the light rays back through the hole. Since catadioptrics have closed tubes, image-degrading currents are eliminated and the optics are basically maintenance free.

Because of the folding of the optical path the instrument is extremely portable and compact. Like the Newtonian reflector, it is free of chromatic aberration. Hence catadioptric (Schmidt-Cassegrain/Maksutov-Cassegrain) instruments combine the best elements of refracting and reflecting telescopes and are, for many, the most advantageous of all.

Oculars (eyepieces)

The other optical part of the visual telescope, the eyepiece, is nearly as important as the objective, as it enables the observer to magnify the object observed. Eyepieces are not generally a fixed part of the telescope, but are interchangeable. In this way a range of magnifications can be obtained.

The greatest magnification obtainable with a given telescope depends upon the quality of the objective lens or mirror, the quality of the eyepiece, the telescope mount, and the state of the atmosphere, or 'seeing' conditions.

A mistake commonly made by beginners is to use high magnification in an effort to distinguish more detail. In fact, it is preferable to use the lowest power that gives the necessary enlargement. For most work, a magnification of about 30-40 for each 25 mm of aperture is sufficient.

Depending on the atmospheric conditions and the quality of the telescope and eyepiece, one can magnify only to a certain point, beyond which the image is not clear or is too hard to keep in the field. Even under the most ideal conditions one does not need the highest magnification. A certain portion of light to size in the image is essential for distinctness; and although by using a stronger eyepiece one can easily enlarge the size, one cannot increase the light as long as the aperture is unchanged. Higher magnification exposes the inherent imperfections of the telescope. The atmosphere becomes more visible and the image becomes dim and indistinct.

Eyepieces or oculars are compound magnifiers, and their function is to magnify the primary image formed by the objective lens or primary mirror of the telescope. Their size is marked in focal length.

To calculate the magnification of a particular eyepiece, divide the focal length of the eyepiece into the focal length of the main telescope.

$$\textbf{Magnification (power)} = \frac{\textbf{1 000 mm}}{\textbf{20 mm}} = \textbf{50 x}$$

Remember that the greater the magnifying power of the eyepiece, the smaller the field of view, or area of the sky, that one can see. No telescope will show a star as anything but a dot or pin-point of light. If you look at a star and see it as a large disc, you may be absolutely sure that there is something very wrong with the telescope, or that the telescope is not in focus!

Viewfinder

For astronomical telescopes it is necessary to have a viewfinder or finderscope attached to the telescope's optical tube. The viewfinder is a miniature lower-power instrument, in general magnifying about 5-8 times, with which to locate objects before viewing them through the main telescope.

The finderscope/viewfinder requires alignment, or collimation, with the main telescope. This is best done during the day. Using a low-power eyepiece, point the main telescope at any easy-to-locate land object (the top of a telephone pole or TV aerial) some distance away. Centre the object in the telescope and adjust the viewfinder's crosshairs until they are exactly on the object already centred in the main telescope.

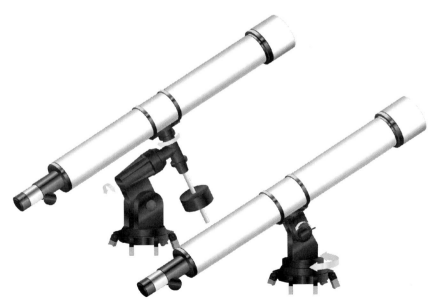

Frank Flowers

Left, equatorial mount. Right, altazimuth mount

Telescope mounts

The basic requirement of an astronomical telescope mount is that it should enable the instrument to be turned to any position in the sky and that it should have rock-like stability combined with smoothness of action that will allow small adjustments of position to be made with precision.

The altazimuth mount is probably the simplest form of mount and has been traditional for small refracting telescopes for many years. For serious observational work it only has application for comet seeking and other specialised work. The altazimuth mount turns on two axes, both of which must be rotated manually in order to follow the motion of the stars and planets.

The type of mount generally used for all forms of observation is one which has its main axis parallel to the Earth's axis of rotation. The basic advantage of the equatorial mount is that it holds an object in the field of view against the Earth's rotation. The telescope is only required to move in this main axis, and the movement should be made at a constant angular speed – a feature which is very convenient for photography. Once the telescope's polar axis has been pointed to the South Celestial Pole, only one axis needs to rotate to follow the motion of the planets, Moon, stars, and so on.

Computer-controlled telescopes

The major manufacturers of astronomical telescopes now produce instruments designed for the amateur that have a built-in microprocessor control system and catalogue with objects stored in memory. With these telescopes you no longer have to 'know your sky' to enjoy astronomy. The onboard computer locates objects for you and tracks them during the viewing session. These telescopes are mounted in altazimuth mode: once the instrument is pointed to the north and you have typed your coordinates into the hand controller (latitude, longitude, time, etc.), the computer does all the calculation required. But it is still fun to learn the night sky!

Bright Stars
and
Constellations

Looking at the night sky we initially see a jumble of stars which, however, seem to sort themselves out into groups or constellations. The form of each constellation is merely a result of the accidental arrangement of stars in space, and has no fundamental significance. The ancient observers saw in these haphazard groups the figures of gods and animals, and a folklore has grown up around the constellations. Without some knowledge of mythology many of the great works in literature cannot be appreciated and understood. The same applies to the constellations and their names. In order to understand why some of the constellations are grouped together, it is necessary to acquaint oneself with the ideas of the structure of the universe which prevailed among the Greeks – the people from whom we received our science.

Most of the ancient observers of the sky lived in the northern hemisphere, so that we, south of the equator, see the heavens reversed. For example, the constellation of Orion depicts a man standing on his head, while the familiar winter constellation Scorpius (The Scorpion) signals the centre of our Milky Way with a downward-pointing tail. A large portion of the southern constellations was first named about two centuries ago by a French astronomer, Lacaille, while on an expedition to the Cape of Good Hope.

Eighty-eight constellation boundaries encompassing both hemispheres were officially sanctioned in 1928 by the International Astronomical Union, a worldwide congress of professional astronomers. All the stars within a particular boundary are members of the same constellation regardless of whether or not they are incorporated into the pattern. The constellation 'pictures' have not been standardised, which has led to a great deal of variation. Most modern star atlases simply ignore the problem by only representing the boundaries and not the figures themselves. The unofficial grouping of stars is termed asterism: the Pleiades (Seven Sisters) is an example of a well-known asterism often mistaken for a constellation.

Hubble Heritage Team (AURA/STScI/ NASA)

Stars come in different colours, which indicate their surface temperature. Most of the stars in this picture of the Sagittarius Star Cloud are orange to red and relatively faint, as our Sun would appear. The bright red stars are cool Red Giants. Stars of the Sagittarius Cloud lie towards the centre of our Galaxy.

The reason for standardising the constellation boundaries was not to memorise their ancient lore, but rather as a means of rapidly identifying the region of the sky in which an object of interest is located. Since the constellation regions are large with respect to the position of individual objects, there exist more exact methods for accurately positioning a telescope so that an observer can identify a particular star or object. The coordinates system by astronomers is termed Right Ascension and Declination. It is simply the Earth's longitude and altitude projected into the sky.

The bright stars in each constellation are named – normally in order of decreasing brightness – by the letters of the Greek alphabet. Quite a number of bright stars have proper names, most often of Arabic origin, in addition to their specification by the Greek letter and the constellation name. Therefore, the brightest star in the constellation of Scorpius is Alpha Scorpii or α Scorpii, but it also has the name Antares.

Once the 24 characters of the Greek alphabet were exhausted, Johannes Bayer repeated the procedure using the letters of the Roman alphabet, first lower case, then upper. Johan Flamsteed devised an alternative method by which the positions of stars of any constellation were numbered, starting from their westernmost boundaries, in order of the increasing eastward position or right ascension. Today, it is common practice to list the brightest stars of a constellation with Bayer's Greek letter (lower case) and the fainter stars by their Flamsteed numbers. Thus, one rather famous star of modern times is known as 61 Cygni. It is a faint star in the constellation of Cygnus (The Swan), the first star whose distance was determined by the German astronomer, F W Bessel. (In fact, the first attempt to measure the distance of a star – Alpha Centauri – was made in the 1830s by Thomas Henderson during his short stay in Cape Town as the Director of the then Cape Observatory. Unfortunately, by the time Henderson returned to his beloved Scotland, Bessel had published his results in Germany, so, officially, the honour goes to Bessel.)

Star Magnitudes

Some 9 000 stars can be seen with the unaided eye from all over the Earth throughout the year, but only about 2 500 to 3 000 can be seen at any one time in any one place. These range from the sixth magnitude to the first. The system of magnitude that astronomers use to denote the brightness of stars was invented in ancient Greece by the astronomer Hipparchus. It was he who devised the terminology – calling the brightest stars he saw first-magnitude stars; those about one-half as bright, second-magnitude stars, and so on to sixth-magnitude stars, the dimmest he could see. When the telescope came into use, astronomers extended Hipparchus's magnitude scale to larger magnitudes to describe the dimmer stars visible through their instruments.

New techniques have since been developed for measuring more exactly the amount of light energy arriving from a star. The branch of science dealing with such measurements is called photometry. Astronomers set out to define the magnitude scale more precisely. Their measurement showed that a first-magnitude star is about 100 times as bright as a sixth-magnitude star. In other words, it would take 100 stars of magnitude +6 to provide as much light energy as we receive from a single star of magnitude +1. Therefore, the magnitude scale was redefined so that a magnitude difference of 5 corresponds exactly to a factor of 100 in the amount of light energy received, so it takes about 2.5 third-magnitude stars to provide as much light as we receive from a single second-magnitude star.

There are a few stars brighter than first-magnitude. These are reckoned on the same scale. Stars 2.5 times as bright as first-magnitude are designated a zero magnitude. Stars brighter than that are labelled minus. Thus Sirius, our brightest star, is magnitude -1.6.

These magnitudes are properly called apparent magnitudes because they describe how bright an object appears to an Earth-based observer. Apparent magnitude is a measure of the energy arriving at the Earth, and does not measure the actual brightness of the stars. A star that looks dim in the sky might really be a very brilliant star that happens to be extremely far away. In order to determine the actual brightness of a star, the exact distance must first be known.

Measuring Distance

How do astronomers measure the distance of the star clusters, nebulae and other galaxies?

The distances of the nearby stars are measured by the mathematics of the land surveyor. This method of measuring the distance of an unapproachable object depends upon the relationship between the angles and sides of a triangle. If one side of the triangle (the base-line) is known together with the angles made at both ends by the other two sides, the astronomer can compute the distance from the centre of the base-line to an object at the apex of the triangle, where the other two sides meet.

As the distance of objects increases, however, the method of the surveyor, good enough for the surface of the Earth, good enough even for the vast stretches of space within the solar system, becomes increasingly clumsy. It becomes more and more difficult, and finally impossible, to find a base-line long enough. In reaching out to the stars the other two sides of the triangle become so long that the longest base-line available to humans – the diameter of the Earth's orbit – is negligible by comparison.

The Cepheid variable stars, which change their brightness as regularly as clockwork, show a definite relationship between the length of time required for their variations in light and their absolute magnitude. The apparent brightness of a star is the resultant (total outcome) between actual brightness and distance. A small, dim star may be near to us and seem brilliant, while a blazing giant may be so far away that it is almost invisible. If one can reduce the brightness of any star to its real or absolute magnitude – the brilliance it would present at a standard distance of 32.6 light years – one can estimate its distance.

It was possible to measure the distance of some of the nearer Cepheid variables and determine just what their absolute magnitude was. Further, it was developed that every Cepheid variable with the same period possessed the same absolute magnitude. And that was all the astronomers needed. Peering at the spiral nebulae, lost in space at unthinkable distances from Earth, astronomers found that Cepheid variable stars existed in those nebulae. They timed the period of the variables and thus knew the absolute magnitude – the intrinsic brightness – of the object. Then, knowing their real brightness and the brilliance at which they appeared, it was a comparatively easy matter to estimate just how far away they and their spiral nebulae were. In addition, once the distance of the nebulae and the star clusters was known, it was not hard to calculate their size.

Henrietta Leavitt

In 1908 Henrietta Leavitt noted that in one of the Magellanic Clouds there was a strong relationship between the brightness of the Cepheids and their period of pulsation. The period-luminosity, as it is called, has been studied in great detail and is now one of the most important methods used to determine the distance to other galaxies.

So have scientists applied a tape measure to the galaxy and to the space through the universe. They have studied and classified other galaxies outside our own.

Galaxies

Humankind's exploration of the environment has now penetrated to the furthest depths of the universe, almost as far back as the beginning of time itself. With the Hubble Space Telescope and powerful earth-based telescopes the professional astronomer can look out practically to the limits of the cosmos. From these furthest boundaries radiation takes billions of years to journey to our planet, with the result that astronomers are looking back across aeons of history when they survey the distant universe.

The Hubble Deep Field: Astronomers selected an uncluttered area of the sky in the constellation Ursa Major and pointed the Hubble Space Telescope at a single spot for 10 days, accumulating and combining many separate exposures. This is the result – our most distant optical view of the Universe. The dimmest, some as faint as 30th magnitude (about four billion times fainter than stars visible to the unaided eye), are very distant galaxies and represent what the Universe looked like in the extreme past.

Through our vast universe there are countless isolated star systems, separated by immense distances and differing greatly in size and brightness. Two centuries ago, before their true nature became evident, these starry concentrations were termed nebulae (Latin for clouds), because of their misty appearance when viewed through a simple telescope. Today we refer to the wheeling star systems that make up the 'atoms' of our universe as galaxies.

There are three broad categories of galaxies, devised by Edwin Hubble: spiral (normal and barred), elliptical and irregular. Our own Milky Way system looks like a spiral from the outside.

Spiral galaxies are typified by the famous Great Nebula in Andromeda, M31, (NGC224), NGC 7424 in the constellation of Grus, M104 (NGC 4594) the Sombrero galaxy and M100 (NGC 4321) in the Coma Berenices cluster.

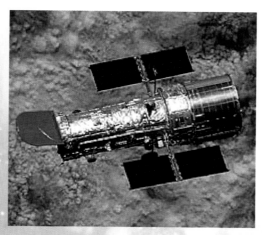

One of the most ambitious experiments in the history of science was the launch of the Hubble Space Telescope (HST) on 24 April 1990. The telescope was named in honour of the American astronomer, Edwin Hubble, who was the first to prove that the objects once called spiral nebulae are, in fact, independent star-systems. The HST is a reflector telescope with a 2,4 m mirror; it is 13 m long and weighs 11 000 kg. It was launched in the Space Shuttle Discovery and put into a near-circular orbit which takes it round the Earth in a period of 94 minutes at a distance of almost 600 km. It is an American controlled project with strong support from the European Space Agency (ESA).

NGC 7424

Within the Virgo Cluster of Galaxies the Hubble Space Telescope captured what is known as 'The Mice' (NGC 4676) because of their long tails. They are situated about 300 million light years away.

Elliptical galaxies are very common. Their name is derived from their smooth, elliptical shape. Unlike the spiral galaxies they have little interstellar material left, so that star formation in them has practically ceased. Typical examples of elliptical galaxies are NGC 2787 in the Virgo cluster of galaxies and M89 (NGC 4552).

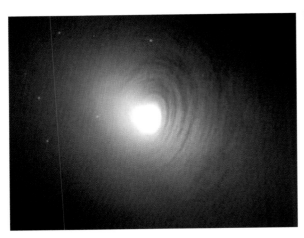

NGC 2787

'Irregular' is really a catch-all name for those galaxies that do not fit the spiral or elliptical nomenclature. These are distorted or irregular in outline as can be seen in the photograph of NGC 1427A taken by the Hubble Space Telescope. Over 20 000 light years long, it is moving towards the Fornax cluster of galaxies.

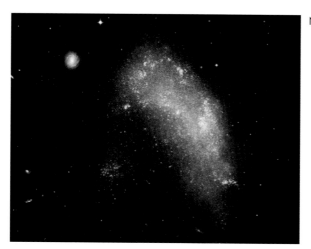

NGC 1427A

Optical, radio and infrared astronomers have shown that, in addition to stars, galaxies contain considerable amounts of gas and dust. The proportion of gas and dust compared to the total mass of a galaxy varies with the type, but it is generally highest for the irregular and the lowest for elliptical. By far the most common element in the gas is hydrogen. Beyond the main system there is the galactic halo, which is more or less spherical, and contains objects which are very old, such as globular clusters and highly-evolved stars.

The Milky Way

On a clear, moonless night away from city lights, a heavy band of brightness stretches across the sky. This hazy band is actually our inside view of the Milky Way – a vast, disc-shaped assembly of several hundred million stars that includes our own Sun.

Our Galaxy is about 100 000 light years in diameter and, on average, about 2 000 light years thick. Spiral arms suggest that our Milky Way, like other spiral galaxies, rotates. The Sun lies between 25 000 and 30 000 light years from the centre, not far from the main plane and near the edge of the spiral arm. This is the arm we see during our winter months when we look at the portion of the Milky Way stretching across Scorpius and Sagittarius. We cannot see through to the centre of the galaxy, because there is too much obscuring material in the way.

Sweeping the Milky Way with binoculars or a low-power telescope will reveal so many stars that to count them in an ordinary way would take more than a lifetime. The difference in star density within the Milky Way, as compared with its outer regions, is very noticeable. The plane of the Milky Way contains opaque dust and gas as well as stars, which gives it its patchy appearance.

The chief constellations through which the Milky Way passes are Aquila, Sagittarius, Scorpius, Centaurus, Vela, Puppis, Monoceros, Orion, Taurus and Auriga, as seen from a southern hemisphere aspect.

Portion of the Milky Way

Johannesburg Planetarium

Our solar system seems completely isolated. The nearest star, a member of the local star cloud to which the Sun belongs, is about four light years away. Around the solar system on all sides stretches space that is far more void of any matter than the most perfect vacuum obtainable in our laboratories; that has virtually no temperature at all; that is pervaded throughout by absolute zero. Black and soundless, this space engulfs the solar system. It surrounds everything in the Milky Way galaxy, in fact, and stretches beyond, setting it off from other galaxies just as completely as the Sun is set off from the stars.

In the Milky Way universe, our galactic system, there are millions of stars like the Sun – some thousands of times larger, some smaller. Some are gathered by the thousands in globular clusters, some travel together in larger groups comprising other clusters, some have one or two companions in a closely knit system, and some travel singly along their endless paths.

Among them are the few ring nebulae, hollow spheres of gaseous material; diffuse nebulae, looking almost like puffs of cotton wool set against the blackness of space; and the dark nebulae. Both bright and dark nebulae are believed to be clouds of cosmic dust composed of extremely tiny particles. Such material is probably scattered very sparsely throughout otherwise empty space. Some of the particles, congregating because of various influences, form vast clouds which probably blot out great sky regions. In their dark form they hide immense sections of the heavens, resulting in what seem to be black holes among the stars. Diffuse nebulae are illuminated by reflection from nearby stars.

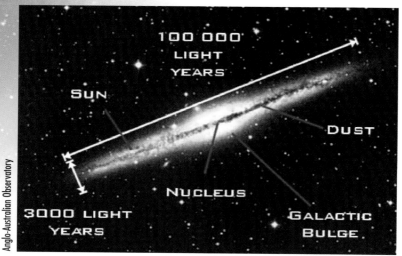

Anglo-Australian Observatory

Diagram of the Milky Way: This image shows a galaxy similar to our own, with the Sun in the position it would be in our Milky Way – about two-thirds of the way to the edge.

Even clouds of cosmic dust have their own motion, as do all the other units of the universe. Each planet speeds along its own path, its satellites circling around it; comets, meteors and asteroids wheel about – all in the retinue of the Sun. All the stars are rushing through empty space, and so are star clusters and the ring nebulae, speeding towards some unknown destination. At the same time the entire galaxy is turning at an immense speed about a central axis located in the direction of the constellation Sagittarius. It requires something in the order of 225 million years for the system to rotate once upon its axis.

Galex team, Caltech, NASA

Viewed in ultraviolet light, the M31 Andromeda Galaxy, the closest major galaxy to our Milky Way looks more like a ring galaxy than a spiral. M31 lies about three million light years distant and is bright enough to be seen without binoculars, provided one knows where to look. It is, however, not well placed for Southern Hemisphere observers as it is low above the northern horizon and situated in the constellation of Andromeda. This mosaic image of M31 shows young blue stars dominating the picture.

The galactic system has this motion in common with the other universes – the spiral nebulae – that have been observed. Similar in many respects to the Milky Way system, these universes are situated at great distances from us, and there are millions of them. One of the largest of these 'island universes' is the great Andromeda Nebula, which is some 2 million light years away.

The spiral nebulae contain stars, star clusters, variable stars, novae, and great clouds of the same type of cosmic dust as occur in our own system. They, too, seem to rotate about a central axis, although the short time during which they have been under observation precludes our seeing any changes.

Large Magellanic Cloud

Our Milky Way can be classed as just another of the spiral galaxies that lie about in space. So much is it one of them that, together with neighbouring universes, it comprises what astronomers know as a local super-galaxy. The individual members include the Milky Way system, with its millions of stars; the Magellanic Clouds, which are very close companions of the Milky Way; the Great Spiral galaxy in the constellation of Andromeda; M33, together with its companion universes; and probably others that are obscured by dark cosmic dust clouds in the Milky Way.

There are other super-galaxies in space – the Coma-Virgo group, for instance – which contain many more units than this. Each of the separate galaxies comprises a universe unto itself; with their companions, they become members of a super-universe, bonded together by their location in space. By what ties they are united, of what possibly larger unit they may in turn be members, we have yet to learn.

Novae

Literally, these objects are termed 'new stars', although this is a bit of a misnomer.

A nova is in fact a double star system in which one star is 'normal' and the other is a white dwarf. Matter accretes in a thin layer on the surface of the white dwarf and eventually ignites in a thermonuclear explosion. This blows a thin surface layer off into space, causing a larger rise in light output from the system.

Occasionally there occurs a supernova, a vast explosion in which an entire star is destroyed. These are mostly seen in distant galaxies as a 'new' star, and are extremely bright for a few days, rivalling the combined light output of all the rest of the stars in the galaxy. The brightest supernova seen in our galaxy, numbered 1987A, occurred in the Large Magellanic Cloud, the small satellite galaxy to the Milky Way, visible from the southern hemisphere.

Supernova remnant

Supernovae fall into two different types. Type I results from mass transfer inside a binary system consisting of a white dwarf star and an evolving giant star. Type II supernovae are, in general, single massive stars which come to the end of their lives in a very spectacular fashion.

Within stars there is a constant battle between gravity and radiation pressure rising from internal energy generation. In the early stages of a star's evolution the energy generation at its core comes from the conversion of hydrogen into helium. For stars with masses of about 10 times that of the Sun this will continue for approximately 10 million years. After this the hydrogen at the centre of the star is exhausted and hydrogen 'burning' can only continue in the shell around the helium core. The core contracts under gravity until its temperature is high enough for helium 'burning' and conversion into carbon and oxygen to occur. This phase also lasts about a million years, until the helium at the star's centre is exhausted. The core again contracts until it is hot enough for the conversion of carbon into neon, sodium and magnesium. The crunch comes when the mass of the star reaches 1.4 solar masses. Gravity compression heats the core, and the core collapses. This releases an enormous amount of energy and later a shock wave. When this shock wave reaches the star's surface the temperature rises to 20 000 degrees and the star explodes. In Type II supernovae the central neutron star remains.

Double and Multiple Stars

Two tiny points of light, one a rich orange, the other deep blue, placed close together in the telescope field of view – such is the appearance of Albireo, the eye of Cygnus, and certainly one of the most spectacular examples of a double star visible in the southern hemisphere night sky. The concealed beauty of many similar stellar objects lies unsuspected until discovered in the telescope. Look for them if you can; coloured doubles are the jewels of the sky.

A surprisingly large number of doubles lie within reach of the amateur astronomer's telescope. (A double star simply means a star that is single to the naked eye but can be resolved into two stars with a telescope.) Double stars can be found in almost any quarter of the sky and offer various combinations of colour, magnitude and other features. An observer can become familiar with as many as desired, and once learned doubles seem like old friends when re-observed from time to time.

There are also complex systems of more than two stars, known as multiple stars. Many examples belong to more than one of these groups – they may double in a small telescope but prove to be triple or multiple in a larger one. A good example of a multiple star is the quadruple theta Orionis, whose components form the trapezium in the Great Nebula of Orion. The stars range from 4.7 to 8 in magnitude and are white, lilac, garnet, and reddish in colour. The most famous of all multiple stars in the southern hemisphere is alpha Centauri – a triple star system – best known as the star nearest to our Sun.

To get the best effect when observing double stars, one should use magnifying power with care – it is best to use the lowest power that will resolve the pair nicely. The essentials for double star work are a list of objects and an atlas by which to locate them. If you appreciate the beauty of colour you will soon become enthusiastic about the wide range of colours and the often striking contrasts in the various doubles. However, differences in the colour correction and apertures of telescope lenses or mirrors and the variations in different eyes and even in the atmosphere mean that reports of colour in doubles have always been notoriously discordant. As a result colours are recorded in different ways in different star lists. In some stars the hues are very distinct and give no difficulty; in others they are particularly elusive.

The closest star to the Sun is the Alpha Centauri system. Of the three stars in the system, the dimmest – called Proxima Centauri – is actually the nearest star. The bright stars, Alpha Centauri A and B, form a close binary. In this picture the brightness of the stars overwhelms the photograph, giving an illusion of great size, even though the stars are just small points of light. Alpha Centauri, also known as Rigil Kentaurus, is the brightest star in the constellation of Centaurus and the fourth brightest in the night sky.

1-Meter Schmidt Telescope, ESO

The significance of colour in stars relates not only to their intrinsic beauty but also to the fact that these bodies represent other suns at various stages of temperature and evolution, with possible planets attending them, just as our Sun represents the centre of our solar system. The only sun we know is a yellow-white object.

The observer's training is something of a factor, too, for the real beginner does not often state the colour, even of Albireo, correctly – which seems incredible to the trained telescope user. Probably the faint pastel tints showing traces of violet, or blue-green, or ashen, are the most difficult. Whatever the case, the stars should be kept in the middle of the field of view of the telescope to minimise colour aberration in the lens.

Star Maps

The star maps provided on the following pages are for latitude 26° South (Johannesburg), but can be used anywhere else in the country. These maps are not precisely accurate, and on a flat projection there is bound to be a certain amount of distortion; they should, however, serve your immediate needs. Because stars appear to rise about four minutes earlier on each successive night, it follows that every week they will rise about 30 minutes earlier than they did the previous week and every month they will rise about two hours earlier than they did the previous month. This is why we see different stars and constellations throughout the year.

The Evening Sky of Summer (Looking North)
1 Dec - 10pm, 16 Dec - 9pm, 31 Dec - 8pm, 15 Jan - 7pm

Orion (The Mighty Hunter)

Orion is generally considered to be the most beautiful and imposing constellation of the heavens. It is one of the most easily recognised. Four bright stars form a large rectangle, and the three second-magnitude stars (from west to east, named Mintaka, Alnilam, and Alnitak), equally spaced and forming a straight line (the Belt of Orion), enclosed by a triangle, are a delight to the eye. Immediately above the belt, a line of three fainter and fuzzier looking stars form the sword hanging down from the belt.

The brightest star in the constellation is Betelgeuse, alpha (α) Orionis, bright red in colour, depicting the Hunter's right shoulder. The star is relatively cool, and most of its radiation radiates not in visible light, but as light of slightly longer invisible radiation (infra-red radiation). If we could see the radiation (and various instruments are available which can detect it) we should find this to be one of the brightest stars in the whole sky. By various methods it is possible to estimate the true diameter in kilometres of certain stars, one of which is Betelgeuse. Betelgeuse is so large that, were it located where the Sun is, the orbit of the Earth would be inside it. It is at a distance of approximately 310 light years.

The star Bellatrix, gamma (γ) Orionis, marks the position of the Hunter's left shoulder. Although Rigel, the left knee, which is at least 900 light years distant, is designated beta (β) Orionis, it is slightly brighter than Betelgeuse, approximately 310 light years away. Betelgeuse is an unstable, variable star, swelling and shrinking, changing its magnitude as it does so. At times it is almost as bright as Rigel. The star Saiph, kappa (κ) Orionis, depicts the right knee. Orion is in a portion of the sky that contains seven of the 20 brightest stars in the heavens (the others are to be found in Auriga, Gemini, Taurus, Canis Major and Canis Minor).

There are two famous nebulae in Orion. One, the Great Nebula (M42)*, is visible to the naked eye. It is the prototype of the diffuse nebula, a great cloud of cosmic dust approximately 26 light years in diameter and 1 625 light years away. The star theta (θ) Orionis marks the centre of the Great Nebula.

Orion

Francois du Toit

*M42 is the number in Messier's catalogue. In about 1760 Charles Messier tried to be the first to see the return of Halley's Comet. A tail-less comet looks very much like a faint nebulosity, and an observer looking for comets might waste time checking on these nebulosities, unless he or she knew exactly where they were located. In an effort to avoid such mistakes, Messier drew up and published a catalogue of various deep sky objects - star clusters, nebulae, and so on – and gave them numbers. Although Louis XV referred to him as the 'ferret of comets', today he is remembered far less for his comet-spotting abilities than for his famous catalogue, which is still in use.

Viewed through binoculars, the stars seem to be enveloped in a hazy field that marks the nebula's presence. Even in a small telescope, the Great Nebula is an awe-inspiring sight. In the heart of the nebula are four closely spaced stars forming the 'Trapezium'. A dramatic way to appreciate the extent of the nebulosity is to set the telescope just ahead of the nebula and allow it to drift slowly across the field. In this way the glow of the nebulosity may be traced out far beyond the usual limits. The night should be dark, moonless, and clear, of course, and the eyes well dark-adapted.

The three belt stars serve as valuable pointers – a line through them extended to the south leads to Canopus; to the southeast it points to Sirius, extended northwards it points to Capella, and northwest to Aldebaran in Taurus.

Through the belt of Orion runs the celestial equator. Every star between the equator and the south pole of the sky (south celestial pole) is said to have a declination south (-). Objects between the equator and the other pole of the sky are said to have a declination north (+).

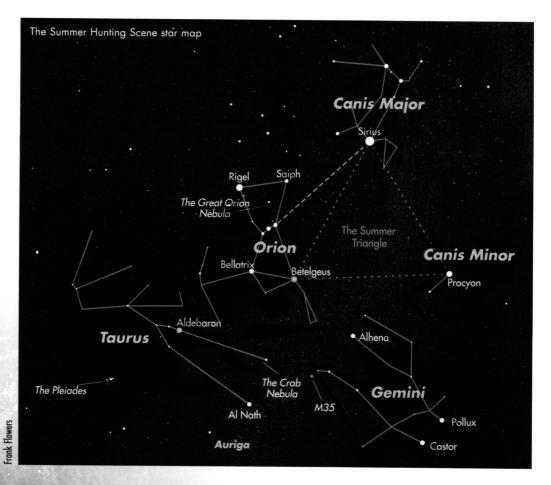

The Summer Hunting Scene star map

Frank Flowers

Canis Major (The Great Dog)

Sirius (α Canis Majoris) is the brightest star in the sky and also the most conspicuous star in the constellation of the Great Dog. It is of a blue colour and is easily located by following upwards (eastwards) the line connecting the three stars in the Belt of Orion. The twinkling of Sirius, especially when it is low above the horizon, is produced by irregular air currents in the Earth's atmosphere. (All stars twinkle more or less, according to the prevailing atmospheric conditions.)

Sirius is the second nearest of the naked eye stars, at a distance of 8.6 light years. It is bright enough to be seen even when it is close to the Sun. The Sun passes Sirius in its passage round the ecliptic early in July. In the northern hemisphere during the heat of summer people called these the 'dog days' because the constellation of Canis Major is seen near the Sun.

Canis Minor (The Little Dog)

Canis Minor is easily located: its brightest star, Procyon, forms an equilateral triangle with Sirius in Canis Major and Betelgeuse in Orion. The name Procyon means 'before the Dog'. The star is so called because it rises about half-an-hour before Sirius, the Dog Star. Procyon, like Sirius, is one of the nearby stars, at a distance of 11.4 light years. The constellation of Canis Minor contains nothing much of interest.

Taurus (The Bull)

Taurus is the second sign of the zodiac and is one of the oldest of the star groups – it was recognised by the ancient Babylonians, Persians, Egyptians, and Greeks. An upside-down 'V'-shaped group of stars, the well-known Hyades cluster, forms the Bull's face. Albebaran, alpha (α) Tauri, marking the Bull's right eye, is a fair-sized red giant star but not an actual super-giant like Betelgeuse. Its distance is 68 light years.

Al Nath, beta (β) Tauri, at the tip of the Bull's exceedingly long left horn, would seem to belong properly to the constellation of Auriga, but is actually shared by both constellations.

Myths of the Constellations – Orion

Orion was the son of Neptune. Hunter of Taurus the bull and Lepus the hare, followed faithfully by his companions Canis Major and Canis Minor, he dominates the northern sky. He was a handsome giant and a mighty hunter. His father gave him the power to wade through the depths of the sea, or, as others say, to walk on its surface. He had no fear of any animal and once threatened to kill all the animals on Earth. When Gaia, the goddess of the Earth, heard this she was furious and sent after Orion the one thing that he would not notice - a scorpion. The scorpion stung him in the heel and the giant fell, mortally wounded, to the ground.

Watching the sky, you will notice that when Orion sets below the western horizon, the stars of Scorpius are just rising in the east. If you watch the following night, anxious to know what happened to the giant, you will see Orion rise again restored to full strength and health by Ophiuchus, the doctor of antiquity who gave Orion an antidote to the poison, saving him from death. When Scorpius sets in the south-west, Ophiuchus stands over him, which indicates that he tramples him underfoot and gives Orion the antidote. And when Ophiuchus sets a little later in the west, Orion comes up in the east fully recovered.

The spectacular Crab Nebula, Messier's No1, is the remains of one of nature's most devastating events, a supernova explosion. This star exploded almost seven millennia ago but the light from the explosion only reached the Earth in the year 1054. The red tendrils are mostly composed of hydrogen gas thrown out into space by the explosion. The blue light is given off by electrons caught in the magnetic field of the exploded star. At the very heart of the nebula is a pulsar – a ball of neutrons that contain more mass than the Sun but squeezed into a sphere with a diameter of about 25 kilometres. This exotic object spins 30 times every second.

The Crab Nebula, a remnant of the supernova seen in 1054, lies close to the star zeta (ζ) Tauri (the horn-tip star below Aldebaran). It can be seen with powerful binoculars, but will look rather dim and uninteresting. However, to astronomers it is one of the most interesting and important objects in the sky. The supernova is now a pulsar, and even now, nearly a thousand years later, the nebula – about 13 light years in diameter – is spinning on its axis at thousands of kilometres per second. It is No. 1 in Messier's catalogue.

The Pleiades (M45)

The Pleiades is one of the brightest star clusters visible during our southern hemisphere summer months. The blue haze surrounding them is fine dust that reflects the blue from the stars. Also known as the 'Seven Sisters', these make up the best known open galactic cluster of stars in the entire sky. They lie in Taurus and are very prominent to the naked eye. People with normal eyesight see at least six of the Pleiades under good conditions; the record exceeds 17, and binoculars will show many more. The total number of stars exceeds 400 and the distance of the cluster is 410 light years. The Pleiades cluster also contains nebulosity. The nebular material shines by reflection and is well studied only when photographed through a large telescope. Look at the Pleiades with binoculars or with a low-power wide-field eyepiece in a telescope. They make a magnificent spectacle!

Myths of the Constellations – Pleiades
The Pleiades were seven daughters of Atlas. One day Orion saw them and, smitten, gave chase. In their distress they prayed to the gods to save them. Zeus snatched them up and placed them in the sky, where Orion still pursues them. The Pleiades are known in Xhosa as Imi-limela – the stars that usher in the ploughing season.

The Pleiades star cluster

Mount Wilson Observatory

Pisces (The Fish)

Pisces lies above Pegasus, but as it contains no bright stars it is not easily recognised. Al Rischa (α Piscium) is the knot in the string to which the two fish are tied. It lies at about 100 light years distance. About 1.5 degrees north-east of α Piscium is a large galaxy, M74. M74 is a fine example of a large face-on spiral galaxy and can be seen in a small telescope of wide field rather than in larger instruments.

Auriga (The Charioteer)

Auriga is a constellation whose four brightest stars form a slightly irregular quadrilateral, with a fifth bright star which, as mentioned, actually also belongs to Taurus, forming a pentagon. The bright star Capella, alpha (α) Aurigae, is readily recognised by its brilliance, its yellow colour, and by the fact that it is flanked by a triangle of fainter stars. Capella, though yellow like our Sun, is a giant rather than a dwarf star – or rather it is two giants, because it has a close companion. One of the stars is 90 times as luminous as the Sun, the other 70 times – the distance between them is not more than about 100 million kilometres.

Capella, the sixth brightest star in the sky, is supposed to be a goat, with kids depicted by the faint stars just below it. The distance from us is 42 light years.

Three interesting galactic clusters for binocular viewing can be found in Auriga – they are M36, M37 and M38. Auriga lies in a very rich region of the Milky Way.

The Evening Sky of Summer (Looking South)
1 Dec - 10pm, 16 Dec - 9pm, 31 Dec - 8pm, 15 Jan - 7pm

Tucana (The Toucan)

Within Tucana can be found the Small Magellanic Cloud, as well as a splendid globular star cluster, 47 Tucanae, which is bright and large but cannot be seen with the naked eye. Use binoculars to locate and view this fine globular, which ranks alongside Omega Centauri. The cluster is about 20 000 light years away.

The star beta (β) Tucanae is a double star system and can be seen quite easily without any optical aid.

The Small Magellanic Cloud is a companion to the Large Magellanic Cloud and both are neighbours to our own Milky Way galaxy.

The Small Magellanic Cloud (SMC) is one of the wonders of our Southern sky at a mere distance of 210 000 light years in the constellation Tucana. Within the SMC is a star-forming region about 200 light years across as pictured by the Hubble Space Telescope. The SMC, a small irregular galaxy, represents a type of galaxy common in the early Universe and thought to be the building blocks for the larger galaxies present today.

ESA, NASA; (A Nota)

Hydra (The Watersnake/Sea Serpent)

A little pentagon of stars forms the head of Hydra, which is the most extensive constellation in the sky. Hydra winds from near Cancer to not far from Centaurus. The spangled serpent stretches a full 120 degrees – one-third of the way around the heavens – yet it has but one bright star – its heart, Cor Hydrae, also named Alphard, (α) alpha Hydra.

Myths of the Constellations – Cancer and Hydra
The second labour of Hercules was to slay Hydra, a many-headed monster. For every head he severed, another two grew in its place. A goddess who disliked Hercules sent a crab to bite his heel while he fought Hydra, but Hercules crushed it under foot. This is why Cancer is the hardest of the constellations in the Zodiac to see. Hydra is right underneath Cancer

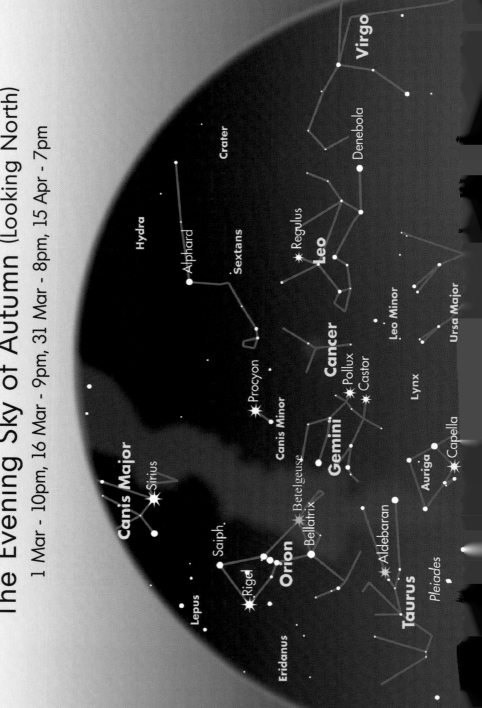

The Evening Sky of Autumn (Looking North)
1 Mar - 10pm, 16 Mar - 9pm, 31 Mar - 8pm, 15 Apr - 7pm

Virgo

Crater

Hydra

Denebola

Alphard

Sextans

Regulus

Leo

Leo Minor

Cancer

Procyon

Canis Minor

Pollux
Castor

Gemini

Ursa Major

Lynx

Canis Major

Sirius

Auriga

Capella

Saiph

Betelgeuse
Bellatrix

Orion

Aldebaran

Taurus

Lepus

Rigel

Eridanus

Pleiades

Gemini (The Twins)

Gemini is the third sign of the zodiac and is made up of two almost parallel lines of stars, with two bright stars, Castor and Pollux, at the head of each line. Castor and Pollux have long been known as the Twins and give the constellation its name. Castor and Pollux form a prominent pair just 4.5 degrees apart. Castor is the northern star and the slightly fainter of the two, shining with a diamond whiteness in contrast to the bright golden tint of Pollux. Pollux is designated beta (β) Geminorum but is actually brighter than Castor, alpha (α) Geminorum. It has been suggested that one of these stars has changed in luminosity in the last few centuries. Castor is 46 light years away from us, Pollux only 36 light years.

M35

In 1781 Sir William Herschel discovered a new world – namely the planet Uranus – beside the star eta (η) Geminorum. And beside delta (δ) Geminorum, in 1930 Clyde Tombaugh finally identified a tiny speck on a photographic plate as the ninth planet in our solar system, Pluto.

The open star cluster M35, considered to be one of the most beautiful in the sky, can be located in Gemini with the aid of binoculars. It is an excellent object for any small telescope, while particularly effective in a large telescope with low magnification. Curving rows of bright stars give the impression of rows of glittering lamps on a chain; fainter stars form a sparkling background, with an orange star near the centre.

M44 Beehive

Cancer (The Crab)

A rather faint constellation, east of Gemini and north of Leo, Cancer contains two Messier objects, M44 and M67. M44 is perhaps better known by the name Beehive Cluster, or its Latin name, Praesepe. It is a bright, open star cluster and on clear moonless evenings it is visible to the naked eye, though it is best viewed with binoculars or a small telescope. Its distance is about 550 light years.

Leo (The Lion)

Leo is among the most ancient of star groups and is quite easily recognised. The stars forming the head of the lion are arranged in the shape of a sickle, or an upside-down question mark, often referred to as The Sickle. The star gamma (γ) Leonis, Algieba, is one of the sky's finest double stars. The yellow and green stars of which it is composed can be seen with a 3-inch (75 mm) telescope. The full constellation includes a triangle of stars to the east forming the hindquarters of the Lion. Denebola, beta (β) Leonis depicts the tuft of the Lion's tail.

Regulus, alpha (α) Leonis, its brightest star, lies almost exactly on the ecliptic and is therefore eclipsed by the Sun once a year (on about 23 August). White in colour and

Myths of the Constellations – Gemini
Castor and Pollux were the offspring of Leda and the Swan, under which guise Jupiter had concealed himself. Leda gave birth to an egg from which sprang the twins. Helen, so famous afterwards as the cause of the Trojan War, was their sister. Castor was famous for taming and managing horses, and Pollux for his skill in boxing. They accompanied the expedition of the Argonauts, and after that we find Castor and Pollux engaged in a war with Idas and Lynceus. Castor was slain, and Pollux, inconsolable about the loss of his brother, besought Jupiter to be permitted to give his own life as a ransom for him. Jupiter consented and rewarded the attachment of the brothers by placing them among the stars as Gemini, the Twins.

130 times as luminous as our Sun, it lies at a distance of 85 light years. The small companion star to Regulus is an easy object for a small telescope. Leo contains well over 70 galaxies visible through a small telescope, but there are no star clusters or nebulae.

Coma Berenices (Queen Berenice's Hair)

An open cluster of stars which should be observed with binoculars on a clear and moonless night when 20 to 30 can be seen, clearly suggesting the shape of a head of flowing hair. There are many star clusters and galaxies in this constellation, hence it is often referred to as the 'Home of galaxies'.

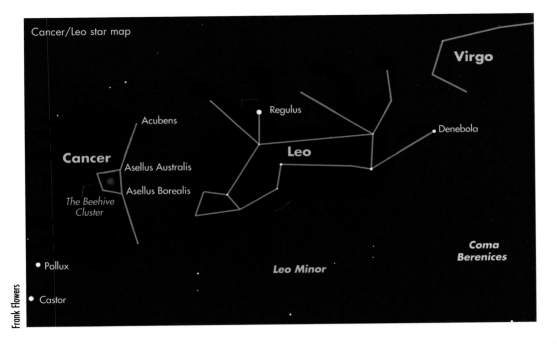

Cancer/Leo star map

Virgo

Regulus

Acubens

Denebola

Cancer

Leo

Asellus Australis

Asellus Borealis

The Beehive Cluster

Coma Berenices

Pollux

Leo Minor

Castor

Frank Flowers

Myths of the Constellations – Leo

Some 4 000 years ago the Sun in the northern hemisphere reached its summer solstice point against the background stars of the Lion. It is easy to appreciate the combination of the regal Sun and the king of the animals. At this time of the year the Sun shines at its hottest in the northern hemisphere as the two kings unite their strength. One Greek story says that this lion lived on the Moon and one day descended to the Earth in the shape of a meteor. It landed in Corinth and there ransacked the countryside until Hercules, as his first labour, strangled it with his bare hands. As such it is known as the Nemean Lion. It has also been said that this group of stars received its name because the Sun's station was here when the early Egyptians watched for the flooding of the Nile. During this season the lions came down to drink, so a likeness of a lion was pictured among the stars.

The Evening Sky of Autumn (Looking South)
1 Mar - 10pm, 16 Mar - 9pm, 31 Mar - 8pm, 15 Apr - 7pm

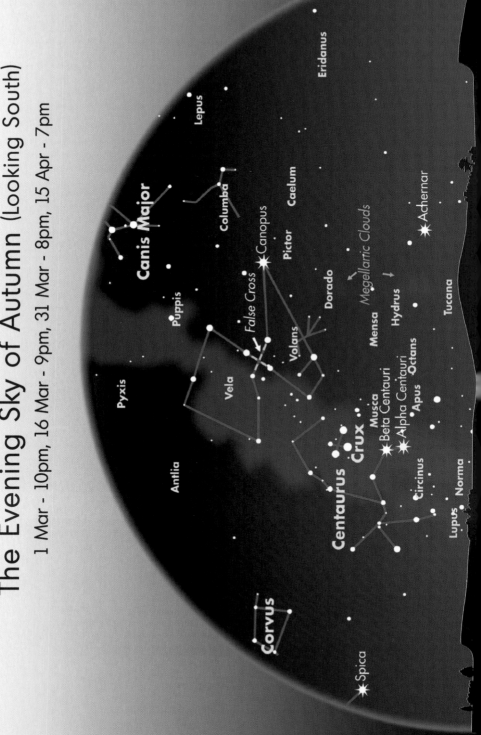

Corvus (The Crow)

Corvus is easy to identify because of its four main stars making up a quadrilateral which stands out because there are no other bright stars in that region of the sky. Look towards the end of Leo's tail, and a little further, close to Virgo, there perches Corvus, the Crow. It is a convenient small group to know. If you cannot locate the Southern Cross (Crux), just find Corvus, and Crux is directly south of it. However, don't confuse Corvus with the Southern Cross. Both are in the shape of a 'kite' but, unlike the Southern Cross, Corvus has no fifth star embedded on the western side, and no two pointers.

False Cross

This is made up of four bright stars. Unwary viewers have often mistaken it for the real Southern Cross. This is understandable because the shapes are the same, but the False Cross is larger, more symmetrical, and less brilliant. During autumn/winter the southern sky is dominated by the star Canopus in Argo Navis, which is much too bright to be overlooked.

Myths of the Constellations – Corvus

Legend has it that Apollo fell in love with Coronis, the mother of the doctor, Aesculapius, and sent the Crow to watch her and report on her behaviour. Corvus had white feathers and a beautiful singing voice. Despite the fact that the report was unfavourable, Apollo placed the Crow in the sky as his reward, but he lost his singing voice and the white feathers were made black.

Another legend says that Apollo sent the raven down to Earth to fetch a cup of fresh water. Raven stayed to eat some ripening figs, and fearing the god's wrath, blamed a serpent for the long delay. Apollo was so angry with the bird that he flung him, cup and serpent out of Heaven. Today we see them together in the sky as Crater (the Cup), and Corvus (the Raven), perched on the back of the serpent (Serpens).

The Evening Sky of Winter (Looking North)

31 May - 10pm, 15 Jun - 9pm, 30 Jun - 8pm, 15 Jul - 7pm

Virgo (The Maid/Virgin)

A rather large member of the zodiac group, Virgo is shaped like an inverted, roughly drawn letter 'Y'. The brightest star is Spica. At a distance of about 230 light years, Spica itself is sufficiently bright and solitary to be recognised at once. Virgo is usually depicted with flowing robes and with an ear of corn in her hand, represented by Spica.

Virgo contains no fewer than eleven Messier objects, including the giant elliptical galaxy, M87, and the remarkable M104, which earns its name of 'the Sombrero'. Unfortunately, a large telescope is needed to see any of these galaxies well.

The Hubble Space Telescope trained its razor-sharp eye on one of the universe's most stately and photogenic galaxies, M104, the Sombrero. The galaxy appears as a brilliant white, bulbous core encircled by the thick dust lanes comprising its spiral structure. As seen from Earth, the galaxy is tilted nearly edge-on. It is ideally placed for Southern Hemisphere observers. The galaxy was named Sombrero because of its resemblance to the broad brimmed, high-topped Mexican hat.

NASA, Hubble Space Telescope

Libra (The Scales)

The four stars that make up Libra at one time belonged to Scorpius. The Greeks knew them as the Claws of the Scorpion. The constellation is almost a square or diamond, between Scorpius and Virgo. Alpha Librae is a double star which, looked at through binoculars, can quite easily be separated.

Libra is a rather obscure constellation and is only noteworthy because it lies in the zodiac.

> **Myths of the Constellations – Virgo**
> Virgo is at times identified as Ceres, the goddess of the Fields and Growing Crops. Sometimes, however, she is also personified as Proserpine, Ceres's beautiful daughter, who was abducted by Pluto, the god of the Underworld. Ceres, pining for her daughter, decided that she would have nothing to do with crops and fruits as long as Proserpine was held captive in the Underworld. Thereafter, everything died, nothing would grow, and the Earth was threatened with famine. Jupiter could not tolerate this, and demanded that Pluto allow Proserpine to spend six months of every year with her mother in the upper world.

The Evening Sky of Winter (Looking South)
31 May - 10pm, 15 Jun - 9pm, 30 Jun - 8pm, 15 Jul - 7pm

The principal constellations are Carina, Crux (the Southern Cross) and Centaurus (the Centaur), Serpens and Argo Navis.

Both Argo and Serpens have changed as the body of astronomy of the Astronomical Union has had to create a rational system from the profusion of stars. All constellations now have borders, and Argo, the celestial ship, a sprawling mass of bright stars, was divided into Puppis, the poop, Vela, the sails, and Carina, the keel. This has resulted in some confusion since the Greek letters employed referred to the old constellation Argo and not to the separate parts into which it is now divided. Serpens has been split into two entirely separate parts – Serpens Caput (the Serpent's head), and Serpens Cauda (the Serpent's tail.)

Centaurus (The Centaur)

Astronomers have used NASA's Hubble Space Telescope to peer into the centre of the dense swarm of stars called Omega Centauri, located some 17 000 light years from Earth. Omega Centauri is a massive globular cluster, containing several million stars. The vast majority of stars in this Hubble image are faint, yellow-white dwarf stars similar to our Sun. It is one of the few globular clusters that can be seen with the unaided eye, and has often been mistaken for a comet.

Without doubt, Centaurus is one of the most splendid groups in the sky, and contains several bright stars quite apart from its two leaders alpha and beta (α and β) Centauri. In shape it is distinctive – it seems to straddle the Southern Cross and does indeed give a vague impression of some large semi-human figure.

Omega (ω) Centauri is the finest example of a globular star cluster in the heavens and one of the most magnificent objects within the range of a telescope. Globular clusters are of great interest to astronomers studying the problems of stellar evolution, since these clusters are known to be the most ancient groups of stars yet identified in our galaxy.

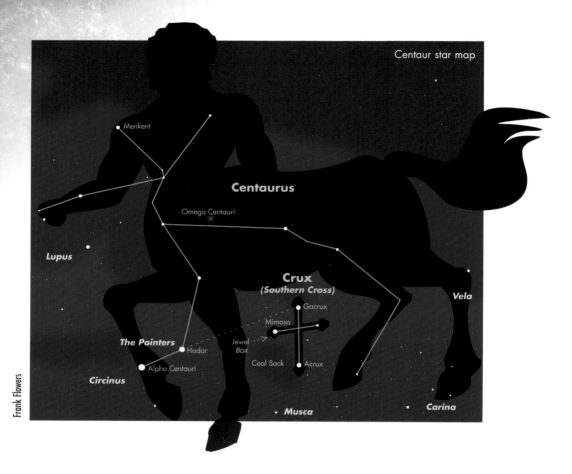

Centaur star map

Menkent

Centaurus

Omega Centauri

Lupus

Crux
(Southern Cross)

Vela

Gacrux

Mimosa

The Pointers

Jewel
Box

Hadar

Coal Sack

Acrux

Alpha Centauri

Circinus

Musca

Carina

Frank Flowers

Alpha (α) Centauri (Rigel Kentaurus or Tolliman) is the third brightest star in the sky. It is a triple star system and famous as the nearest star to our Sun – approximately 4.34 light years. It is also one of the finest visual double stars (binaries) in the heavens.

Beta (β) Centauri (Hadar or Agena) is the eleventh brightest star in the sky. Alpha and beta Centauri, about 4 degrees apart, point towards the Southern Cross.

During our winter months the Southern Cross can be found by identifying four stars of almost equal brilliance. Alpha Crucis is at the foot of the figure, nearest the South Pole; gamma (γ) is at the top, with beta (β) and delta (δ) in the arms. The distance from the top to the bottom of the Southern Cross measures 6 degrees.

South-east of Alpha Crucis is the well-known Coal Sack Nebula. At first glance it looks like a black hole in the sky, but there are a few stars in front of it, though not many. The Coal Sack is only 500 light years away. The San called it the 'Old Bag'. It is a great dark cloud of dust and gas that absorbs the light of the stars beyond it.

Near Kappa Crucis is the 'Jewel Box'. It is about 7 700 light years away, and the stars are thought to be young – no more than a few million years old. Binoculars can be

Don Pettit, International Space Station 6, NASA

The International Space Station (ISS) took this photograph of the Southern Cross. To the left of the photograph are the four stars that mark the boundaries. At the lower left is the dark Coal Sack Nebula, and the bright nebula on the far right is the Carina Nebula.

used to view this pretty group of stars, and an interesting exercise is to try to establish the different colours of the stars within it.

Two interesting features of our southern skies bear the name of Magellan, the great Portuguese seafarer/explorer (circa 1519). Two hazy patches like detached portions of the Milky Way can be seen on a clear winter evening, away from city lights. There is nothing like them in the northern skies. The Clouds of Magellan are island universes (galaxies) and close neighbours of the Milky Way. The Small Cloud is 210 000 light years distant, and lies within the constellation of Mensa (Table Mountain, named after Cape Town's mountain, where La Caille had his observatory); the Large Cloud is about 179 000 light years away, situated within the constellation Dorado (the Swordfish).

A supernova flared up in the Large Magellanic Cloud in 1987 and remained visible to the naked eye for a few weeks.

Myths of the Constellations – Sagittarius and Centaurus
The centaurs were a race of wild creatures, half horse and half man, always armed with bows and arrows. (Sagittarius is sometimes called Centaur, but Centaur is never called Sagittarius.) Centaurs were admitted to the company of man, but on one occasion two became intoxicated with wine and a dreadful conflict arose in which one of them was slain. Unlike his fellow centaurs, Chiron was revered for his wisdom and knowledge of medicine. After being accidentally wounded by Hercules he gave away his immortality and was transformed into the constellation Sagittarius.

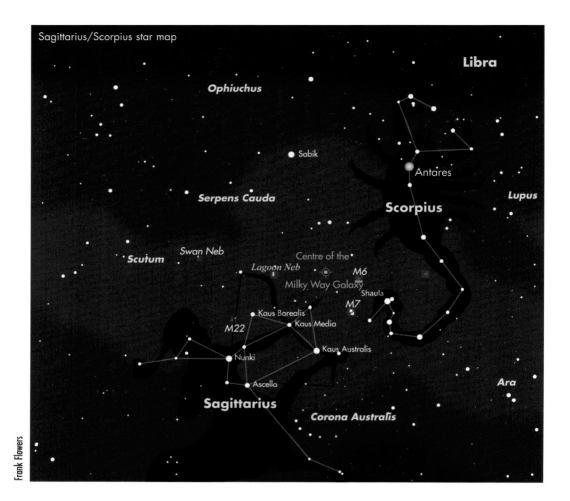

Sagittarius/Scorpius star map

Frank Flowers

Scorpius (The Scorpion)

Scorpius is a very conspicuous zodiacal constellation, and is the most aptly named of all of them, bearing a considerable resemblance to a giant scorpion with its sting poised to strike. The Sun takes only 10 days to move through Scorpius, spending the rest of the 30-day period in Ophiuchus, which is not part of the zodiac.

Antares (from the Greek 'rivalling Mars'), alpha (α) Scorpii, is a super-giant red star, and the brightest in the constellation. It is 7 500 times as luminous as the Sun. If it were in the Sun's position, its diameter (650 million km) would engulf the orbits of the asteroids between Mars and Jupiter. Antares lies at a distance of about 330 light years. But its density is very low – on Earth it would be considered a vacuum. It is one of the four Royal Stars of ancient times, along with Aldebaran, Regulus, and Fomalhaut.

The M6 Butterfly Cluster is located in the constellation Scorpius and is famous for its butterfly pattern. It is best seen at low magnification through a telescope. The entire cluster contains more than 330 stars.

There are many objects worth observing with binoculars in this constellation – it lies in the Milky Way and is rich in countless stars. M6 (The Butterfly) and M7 (which has no name) are open galactic clusters visible to the naked eye under the right conditions and striking when viewed with binoculars. M6 lies about 2 000 light years away.

Sagittarius (The Archer)

Sagittarius is the most southern of the zodiacal constellations, lying between Capricornus in the east and Scorpius in the west. The central part of Sagittarius has a group of stars resembling in shape a giant teapot with a spout and handle, an aid in identifying this constellation. The stars forming the handle and the dome of the teapot comprise a group known as the 'Milk Dipper' – another distinguishing feature.

Sagittarius lies directly in the plane of the brightest part of our Milky Way. The Galactic Centre is located in this constellation. The region is rich in star clusters and nebulae and this is a grand area to scan with binoculars. Observe the dark areas in the Milky Way – these are clouds of cosmic dust with nearby stars to illuminate them.

The famous Lagoon Nebula (M8) is a marvellous diffuse nebula and considered to be a naked-eye object under good 'seeing' conditions. But, while it is easily located with the eyes only, the wealth of detail it contains can only be appreciated with a medium-sized telescope. The Lagoon Nebula is estimated to be at a distance of approximately 5 000 light years.

The Evening Sky of Spring (Looking North)
31 Aug - 10pm, 15 Sept - 9pm, 30 Sept - 8pm, 15 Oct - 7pm

Pegasus (The Winged or Flying Horse)

The feature of this constellation is the so-called Great Square of Pegasus. The bottom right-hand star (eastern end), Alpheratz, actually belongs to the constellation of Andromeda as well, and is designated alpha (α) Andromedae. The square of Pegasus is conspicuous enough, though perhaps not as striking as it looks on a star chart. One interesting experiment may be done: look inside the square and see how many stars can be counted with the naked eye. Then use binoculars and note how many more stars come into view. The difference is quite remarkable. On a clear moonless night, away from city lights, a person with good eyesight can count more than 50 stars within the square. Some of the stars are double, and some grouped in clusters.

Pegasus is one of the few constellations which we in the southern hemisphere see the right way up. In fact it is only half a horse, which perhaps suggests that it is just rising from the ocean foam from which it was born. Once familiar with Pegasus, you can use it as a guide in locating the upside-down 'V'-shaped group of stars of Pisces to the south (see under Evening Sky of Summer, looking north).

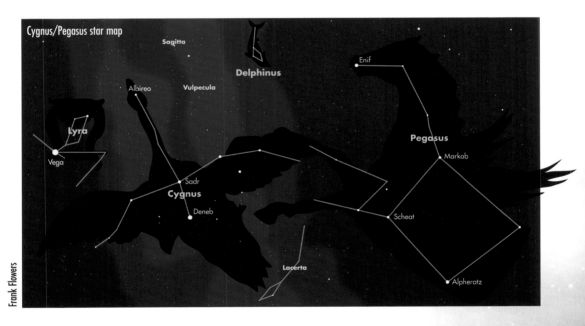

Cygnus/Pegasus star map

Frank Flowers

Delphinus (The Dolphin)

Delphinus is a compact, attractive little group of five rather faint stars. It takes very little imagination to picture a leaping dolphin. This is a splendid area of the sky for sweeping with binoculars as there are many striking and interesting objects.

The Evening Sky of Spring (Looking South)
31 Aug - 10pm, 15 Sept - 9pm, 30 Sept - 8pm, 15 Oct - 7pm

The Evening Sky of Spring (Looking North) continued

Cygnus (The Swan)

Cygnus is a beautiful and easily recognised constellation in the form of a giant cross. It is sometimes called the 'Northern Cross'.

Deneb, alpha (α) Cygni, the brilliant white star of magnitude 1.25, marks the Swan's tail. Deneb is one of the greatest super-giant stars known – at least 70 000 times brighter than our Sun – and is 1 800 light years away.

Albireo, beta (β) Cygni, the eye of the Swan, is one of the most beautiful double stars in the sky, and is considered by many observers to be the finest in the heavens visible through a small telescope and even with good binoculars. The brighter of the two is golden yellow, while the companion is a vivid blue. There are many bright stars in Cygnus – it lies in the galactic plane and is therefore embedded in the Milky Way. Sweep this entire area with binoculars and note the many stars.

Myths of the Constellations – Pegasus
Pegasus was the winged horse born from the blood of Medusa (which sank into the Earth when Perseus cut her head off) and the foam of the sea. Minerva caught and tamed him and presented him to the Muses. Pegasus has always been the symbol of ecstasy and élan. A light-hearted creature, an emissary between the Earth and Olympus, gambolling sometimes in the heavenly fields and at others in earthly planes, he could inspire poets and make heroes perform great deeds.

Grus (The Crane Bird)

Grus lies just below Piscis Australis, and at one time was part of the constellation. It does give a superficial impression of a bird in flight. Grus is rich in galaxies, but these are rather faint and require large telescopes as they are below magnitude 10. It is one of the three southern birds – Grus, Pavo (Peacock) and Apus (Bird of Paradise) – grouped around the South Celestial Pole.

Piscis Australis (The Southern Fish)

Piscis Australis (also known as Piscis Austrinus) lies high in the south-east, along with its brightest star, Fomalhaut. At a distance of 22 light years, it is the nearest of the really bright stars, besides alpha Centauri, Sirius, Procyon, and Altair. Fomalhaut (from the Arabic 'Mouth of the Fish') is about 13 times as luminous as our Sun.

The Southern Fish is seen as lying on its back, drinking water poured from the jars of Aquarius.

Apart from Fomalhaut, Piscis Australis does not contain much of great interest, and most of the stars are fourth to fifth magnitude.

Comets and
Meteors

Tim Cooper

Comets

For centuries we have held comets in awe. In ancient times they were frequently considered to be bad omens; harbingers of bad news and misfortune; the cause of famine and pestilence; heralds of the death of kings or princes; the cause of wars. As an example, comets were blamed for the misfortunes of the Roman emperors Julius Caesar, Claudius and Nero. On the other hand, the emperor Vespasian was unperturbed, stating 'this hairy star does not concern me; it menaces rather the king of the Parthians, for he is hairy, and I am bald'. But they were not always bad news; the fine wine vintage of 1811 was attributed to the appearance of the great comet of that year, and the vintage became known as 'Comet Wine'.

Nowadays, of course, we know that the passage of comets has no effect on man or his environment, though we are still captivated by the appearance of a bright comet in the sky.

Where do comets come from?

Comets are the material left over from the formation of the solar system. The original solar nebula consisted mainly of hydrogen, some helium, and a sprinkling of other elements which, under the influence of gravity, began to form a spherical cloud. The spinning cloud continued to attract material until it became dense enough for nuclear reactions to commence in the centre, and for the Sun to 'switch on'. Much of the remaining dense material coalesced to form the planets and the satellite moons. The gas and dust that was left over was swept up by the Sun and planets or ejected outside the realm of the planets. It is this material that forms the comets that we know exist today. They reside mainly in a huge cloud which has its inner edge just outside the orbit of the planet Neptune. This cloud is called the Oort Cloud, after the astronomer who, in 1950, first suggested its existence. A year later, another astronomer, Gerard Kuiper, theorised that much of the gas and dust left over from the formation of the solar system that failed to coalesce into the planets must exist in the form of a flat disc, still surrounding the planets. Today, this disc is known as the Kuiper Belt and is thought to be a second source of comets.

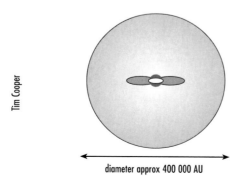

Tim Cooper

diameter approx 400 000 AU

Location of the Oort Cloud and Kuiper Belt. The solar system is represented by the small white disc in the centre. The Kuiper Belt is shown as lobes on either side, surrounded by the spherical Oort Cloud.

In recent years there has been increasing direct evidence of the existence of this disc, with the discovery of faint objects well outside the orbit of Neptune. These objects, known as Kuiper Belt Objects, are thought to be comet nuclei. Similar rings of material have been detected orbiting other stars, the star beta (β) Pictoris being the best known.

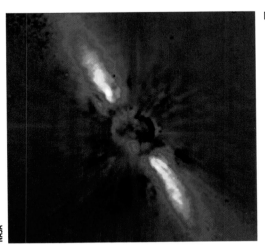

Dust disc around the star Beta Pictoris

NASA

We believe the Oort Cloud contains perhaps a trillion comet nuclei, each in its own orbit around the Sun. Current estimates are that the Kuiper Belt contains about another 200 million comet nuclei. Every now and then a comet nucleus in the Oort Cloud or Kuiper Belt is disturbed, and hurtles inwards under the influence of gravity on a new orbit that carries it close to the Sun.

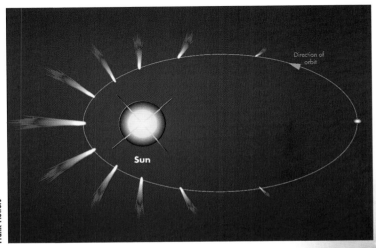

Direction of orbit

Sun

The path or orbit of a comet around the Sun. The diagram shows that the tail of the comet always points away from the Sun.

Frank Flowers

New comets coming in from the Oort Cloud or Kuiper Belt orbit the Sun in a period of thousands of years, and travel in orbits in the shape of a parabola. Very often these long period comets venture close to the Sun or to one of the major planets, especially Jupiter, and the gravitational effect of these large bodies shortens the period and changes the shape of the orbit of the comet. The usual effect is to convert the orbit to an elliptical one. Over a long period, after many returns of the comet around the Sun, the orbit decays, becoming ever shorter.

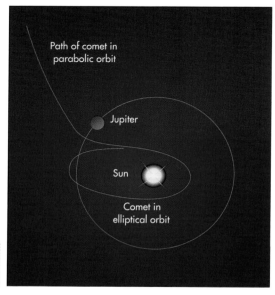

Conversion of an orbit from parabolic to elliptical

Path of comet in parabolic orbit

Jupiter

Sun

Comet in elliptical orbit

Frank Flowers

The anatomy of a comet

When in the frozen depths of deep space, far from the warming effects of the Sun, the comet exists only as a frozen nucleus. The nucleus is made up of the material from which the solar system itself was formed, less the volatile gases, hydrogen and helium, which have long since dissipated. The major constituents are frozen carbon monoxide, carbon dioxide and water, some ammonia and smaller amounts of more exotic materials such as hydrogen sulphide, hydrogen cyanide, formaldehyde, methanol, and so on. So comet nuclei are pretty unappealing places: frigid, smelly and poisonous. Indeed, when the Earth passed through the tail of comet Halley in 1910, a fortune was made by some touting gas masks and comet pills to ward off the predicted effects of the toxic gases in the comet's tail.

In addition to the frozen gases, the nucleus contains a variety of dust particles. The ratio of dust to gas varies from comet to comet depending on where exactly in the solar system the nucleus formed. It was this composition of frozen gases and ices and dust that led astronomer Fred Whipple to refer to a comet nucleus as a 'dirty snowball'.

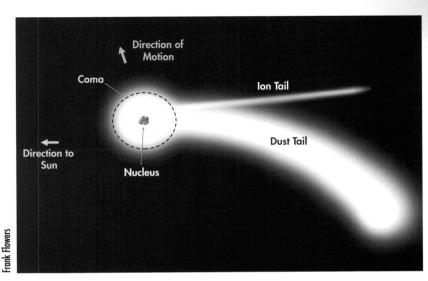

Anatomy of
a comet

As the comet nears the Sun, the frozen gases vaporise under the effect of the Sun's heat. At the same time the bound dust particles are released into the environment of the nucleus. The vaporised gases and ices plus the dust particles form a type of halo around the nucleus called the coma. Usually this coma is so dense that the nucleus can no longer be seen. Visually the comet now starts to look fuzzy, like a tennis ball.

As the comet gets even closer to the Sun it comes under the ever-increasing influence of the solar wind, a continual stream of radiation and charged particles generated by fusion processes going on inside the Sun. The solar wind interacts with the material in the coma, driving it away from the nucleus to form the comet's tail.

Under the correct geometric conditions we can discern two distinct types of tails; one from the gaseous ions emitted from the comet and the other from the dust particles it releases which manifest themselves, in another way, as meteors. The ion tail is formed by ionisation of certain gases emitted from the nucleus by high-energy particles in the solar wind. These particles cause the ions to fluoresce, mainly a blue-green colour owing to the carbon monoxide, as they return to their ground energy state. Because of the repelling of the charged ions, the ion tail appears to point directly away from the Sun, the direction from which the charged particles in the solar wind are coming.

The dust tail, on the other hand, shines by reflecting visible sunlight, so it appears yellowish-white in colour. The dust particles are swept outwards under the influence of the solar wind. The dust tail, unlike the ion tail, often appears curved because of the comet's curved orbit, and the effects of the solar wind on particles of different sizes. It is also possible, though rare, for a comet to display multiple tails. Chésaux's Comet of 1744 showed six or seven tails at a time, pointing upwards from the horizon, while the nucleus was below the horizon.

NASA

Comet Hyakutake

Comet appearances

When can you expect to see a comet? This depends somewhat on the period of its orbit. A long-period comet is one which, by definition, takes more than two hundred years to orbit once around the Sun. In fact few long-period comets take only two hundred years, most of them have periods of thousands of years. For example, the recent Great Comet, known as comet Hale-Bopp, approached the Sun on an orbit with a period of 4 200 years, and must have been seen before, perhaps by the builders of Stonehenge, which dates back to about the same period. Comet Hale-Bopp passed very close to Jupiter in 1996. The effect of this was to shorten its orbit to 2 360 years.

All this means that we cannot predict when the next long-period comet will be seen. About a dozen new ones are discovered each year as they come in towards the Sun and become bright enough to be seen, but most of them cannot be seen with the naked eye. To stand a chance of seeing one of these new long-period comets you need to look out for a discovery announcement.

Under the right conditions a comet can be captured by interaction with one of the giant planets, usually Jupiter, into a short period – less than two hundred years. Most short-period comets have an orbital period below ten years, hence their visibility on a regular basis can be predicted accurately. But because they have passed around the Sun on so many occasions, they have begun to be depleted of their volatile materials and tend to remain quite faint, unless they happen to pass very close to the Earth.

There are a few short-period comets with rather longer periods. Comet Halley is one of them. It passes around the Sun on average every 76 years and usually puts in a

NASA

Hale-Bopp

good performance visible to the naked eye. The comet with the shortest period is Comet Encke, which passes around the Sun once every 3.3 years.

Information about the visibility of comets and periodic comets can be obtained from numerous websites.

How to observe a comet

Firstly you need to know when a comet is expected in the vicinity of the Earth – either the predicted return of a known short-period comet, or a recently discovered long-period comet. Very often, details are published in the press if the comet is bright enough, or can be obtained from the nearest Planetarium or centre of the Astronomical Society. Most comets that are seen each year remain faint and are only visible through binoculars or a telescope. Occasionally, a comet becomes bright enough to be visible through binoculars. Rarely will one become visible to the naked eye or sport a tail.

Remember that comets are notoriously unpredictable, and very often do not become as bright as expected. Do not expect to see any comet from the city or under suburban conditions. Because they are generally faint, fuzzy objects, you must travel out into the countryside to maximise your chances of seeing them. Plan beforehand by plotting the probable position of the comet on a suitable chart, and then hunt it down with binoculars.

Meteors

The solar system is peppered with billions of tonnes of particles, mainly dust grains left over from the formation of the solar system, and material left behind by the passage of comets around the Sun. This inter-planetary dust is the debris of the solar system. In addition to these dust grains, there are thousands upon thousands of small chunks of rock, probably formed through collisions in the past in the asteroid belt. The term given to particles travelling through space is 'meteoroids'. Every now and then the orbit of a meteoroid intersects with the orbit of the Earth, and the meteoroid enters the Earth's atmosphere. The particle is then known as a 'meteor'. Most meteors are too faint to be detected by the naked eye, though they can be located by other means, such as radar, or by scattering radio waves.

The larger dust grains are visible as the 'shooting stars' we often see as we gaze skywards, and in general these grains are about the size of a grain of sand. As the grain size increases, the meteors grow brighter. Any meteor that is equal to or brighter than the planet Venus is termed a 'fireball'. A fireball that is seen to explode is called a 'bolide'. Fireballs are typically the size of a piece of gravel or perhaps a small pebble. Finally, if the object is sufficiently large to reach the ground, it is called a 'meteorite'.

The Earth is bombarded daily by hundreds of tonnes of meteoroids, mostly unseen. But if you stare upwards on a clear night for long enough, you will see at least a handful of meteors. Bright meteors and fireballs must have had a profound effect on the Bushmen – their rock art often depicts images of them. So too, the ancient aboriginals of Australia believed meteors were significant – they saw them as celestial canoes, carrying away the souls of the dead.

Patrick Moore

The Tswaing Crater, 40 km north of Pretoria. A meteorite struck the Earth here some 220 000 years ago, leaving a crater about 1.4 km in diameter and 100 metres deep

There are essentially two categories: sporadic meteors, which are particles travelling through space in isolation; and shower meteors, which travel together through space. Sporadic meteors appear at more or less random times from random points in the sky and travel in random directions. Look upwards on any clear night and you will see perhaps five or six sporadic meteors per hour.

Shower meteors are the debris stream left behind by a comet as it passes around the Sun. They continue to orbit the Sun in an orbit fairly similar to that of the parent comet, and when their orbit intersects that of the Earth, it is possible for a stream of particles to be visible as a meteor shower.

These meteors appear to come from a point in the sky called the radiant, and normally the shower is named for the radiant from which the meteors emanate. For example, the Orionids in October emanate from the constellation of Orion, and the Geminids in December from the constellation of Gemini, and so on.

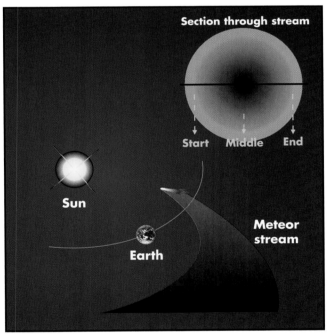

Section through stream

Start   **Middle**   **End**

Sun

Meteor stream

Earth

Path of the Earth through a meteor stream

Frank Flowers

Within the meteor showers, we get the minor showers – which generally show very low activity, below 10 meteors per hour, and the major showers – which have as many as 100 meteors per hour in a good year.

A meteor stream's orbit intersects with the Earth each year on about the same date as the Earth revolves around the Sun in its own orbit. But each year the performance of the shower changes because of differing densities of dust in the meteor stream.

The dotted lines represent meteor paths traced back to the stream radiant. The diagram uses the Southern Cross as a reference point.

Frank Flowers

List of prominent annual meteor showers

Shower	Duration	Maximum	Radiant	Rate
Eta Aquarids	21 Apr-12 May	6 May	22h20, -02	30-50
Scorpius/Sagittarids	June-July	15 June	various	5-10
Delta Aquarids	21 Jul-25 Aug	29 July	22h30, -16	20-30
Alpha Capricornids	15 July-15 Aug	30 July	20h30, -10	5-10
Orionids	2 Oct-7 Nov	21 Oct	06h20, +16	20-30
Taurids	15 Oct-20 Nov	3 Nov	03h20, +14	5-10
Leonids	14 Nov-20 Nov	17 Nov	10h10, +22	10-80
Geminids	4 Dec-16 Dec	13 Dec	07h30, +33	30-50

The table opposite lists only the most important annual showers. A more extensive list appears in the *Handbook* of the Astronomical Society of Southern Africa. The radiant position given is the Right Ascension and Declination of the point from which the shower meteors appear to emanate, for the epoch 2000. The rate is the number of meteors that can be expected at the shower maximum, with the radiant at the zenith and under clear skies where stars of magnitude 6.5 can be seen with the naked eye. Significantly lower numbers will be seen under poor conditions and when the radiant is lower in the sky.

How to observe meteors

If you want to observe a specific meteor shower, check on the dates on which the shower is expected to be active. Best chances are on the night of maximum and for one or two days on either side of maximum. Then check the phase of the Moon. If the Moon is at a few days on either side of New Moon, the conditions are ideal. If the Moon is near full, it will wash out all but the brightest meteors.

Meteor watching is ideal for beginners. You don't need any expensive equipment, and while you look for meteors you can start to identify the constellations. The only equipment you need to get started is a chair or sleeping bag; perhaps a pillow for comfort; a clipboard and pencil with which to record your observations; a watch to record times; and a torch, preferably covered with red cellophane. Red light does not affect adaptation to the dark as much as normal torch light does.

Once you have selected the shower to watch, set yourself up comfortably facing the general direction of the radiant. You can set up on a reclining chair or in a sleeping bag. The latter is excellent during winter to help ward off the cold. Now allow your eyes a few minutes to adapt to the dark. When you are ready to begin your watch, record the time on the clipboard. Observe for a minimum of one hour and count the number of meteors you see in that time. If the meteor can be tracked back to the radiant, then record it as a shower meteor (for example Geminid). If the meteor cannot be tracked back to the radiant, record it as a sporadic. Finally, record the time at which you stop observing. Send your finished counts to your local planetarium or ASSA representative for further processing.

If you enjoy making these observations of meteors, you are ready for some more serious scientific study.

The Sun and the Moon

The Sun

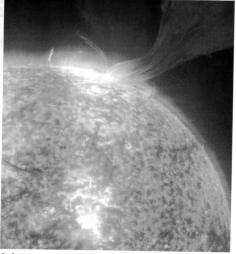

Solar prominences

Basic solar data

Average Earth-Sun distance	49 597 870 km
Maximum Earth-Sun distance	152 100 000 km
Minimum Earth-Sun distance	147 100 000 km
Radius	696 000 km
Escape velocity	617.5 km/s
Mass	1.981×10^{30} kg
Rotation	25 days
Surface temperature	5 800 degrees Kelvin
Core temperature	about 15 000 000 degrees Kelvin
Spectral type	G2
Apparent magnitude	-26.8 (about 600 000 full Moons)
Light-travel time from Sun to Earth	8.3 minutes
Mean distance from centre of the galaxy	25 000 light years
Revolution period around the centre of the galaxy	225 000 000 years

Never look at the Sun with your unaided eye!

Observing the Sun, even for a fraction of a second, can and will cause instant and irreversible eye damage.

The Sun is the nearest example to us of a star, and the only star which appears as a disc and not a mere point of light. It has a diameter of 1 392 000 km and is a huge globe of incandescent gas, mainly hydrogen, with a temperature at the surface of about 6 000 degrees centigrade. It is big enough to hold over one million Earths. Like most normal stars, the Sun creates its own energy by nuclear transformation which takes place deep inside it. At the same time as it releases its energy, it gives away some of its mass. Every second 4 000 000 tons of the star are poured off into space, and so big is the source of life that scientists estimate it can continue at this rate for several billion years.

The Moon

Full moon

Basic lunar data

Diameter	3 476.6 km
	(0.27 of Earth)
Mass	0.012 (Earth = 1)
Mean distance	383 400 km
from Earth	(centre to centre)
Perigee (closest	356 410 km
distance to Earth)	
Apogee (farthest	406 697 km
distance from Earth)	
Sidereal period	27.321661 days
(fixed star period)	
Synodic period (New Moon	29.530588 days
to New Moon)	
Mean orbital speed	3 680 km/h
Orbital inclination	5 degrees 9 min
Escape velocity	2.38 km/sec
Surface gravity	0.165 (Earth = 1)
Mean magnitude	-12.7 (at full phase)
Albedo (ability to reflect	0.07
light from its surface)	

It has been said that we are living on one member of a double-planet system. Mercury and Venus have no moons, and each of the other planets in our solar system is thousands of times more massive than any of its satellites. The Earth is only 81 times more massive than our Moon, implying that the masses of the Earth and Moon are roughly comparable.

There have been three competing hypotheses for the origin of the Moon, all with serious physical problems:

1. The Fission theory, invented by the astronomer George Darwin (son of Charles). He proposed that the Earth spun so fast that a chunk broke off, with the Pacific Ocean as the scar. But this theory is universally discarded today. Firstly, the Moon is chemically different from the Earth; secondly, the Earth could never have spun so fast that it threw a moon into orbit – and if it had, the escaping moon would have been shattered while it was within the Roche Limit (the minimum distance at which a fragile body can survive without being gravitationally disrupted or shattered, named after Edward Roche, a French mathematician).

2. The Capture theory – the Moon was wandering through the Solar System, and was captured by the Earth's gravity. But for an approaching body to enter into orbit around another it would need to lose a lot of energy (which is why spacecraft sent

to orbit other planets are designed with retro-rockets). Otherwise the approaching body would have been 'sling-shotted' rather than captured – the phenomenon the Voyager probes exploited. Also, a successful capture would have resulted in an elongated, comet-like orbit.

3. Condensation (or creation) theory – Earth and Moon formed at about the same time from the same portion of the swarm of planetismals which supposedly orbited the Sun in the early phases of the evolution of our Solar System. However, it is unlikely that the gravitational attraction could have been strong enough.

Planetary scientists will not give up. They will find an answer, but as yet the origin of the Moon is still unresolved.

All the light of our Moon is reflected sunlight; there are no incandescent areas in which it might be shining by its own light. The easiest method of proving this is to look at the dark, invisible limb of the Moon, which is in shadow. When it is not illuminated by earthshine we see nothing of its surface.

The variable line between an illuminated portion and the part in shadow at any moment is known as the terminator. Because the Moon is spheroid, the terminator is actually a complete circle around it, corresponding to the 'circle of illumination' on the Earth. However, this cannot be seen from Earth because the opposite side of the Moon is always invisible to us. Generally speaking, if one was at the terminator any time between New and Full Moon, one would be having sunrise on the Moon. Similarly, if one was on the section which is seen from Earth between Full Moon and New, one would be experiencing sunset.

Our Moon lends itself particularly well to telescopic study because it is so close to the Earth and because of its lack of atmosphere. Nothing on the Moon prevents all the features of its surface from standing out clearly and being sharply defined in our telescopes and photographs. Although only one face of our satellite is open to observation, that face presents a superb array of features to delight the eye. Anyone who has not looked at the Moon with a telescope – or even binoculars – has no idea of the magnificent views, and even the experienced observer of lunar landscapes repeatedly finds new landscapes under new lights or conditions.

Any and all magnifications may be used to view the Moon. Even high-powered binoculars, magnifying 10-20 times, will show the large grey plains. But to do justice to our Moon, one should use a telescope to reveal all the marvellous detail. A 3-4 inch refractor opens up a vast field, while a 6-inch reflector and bigger will reveal the details. Lower magnifications are needed to include the entire disc for general examination. For more detailed observation of mountain ranges and chains of crater-like formations a magnification of 60-100 times is needed, while magnifications of 200-400 can be used for close-ups and details of the Moon's smaller features.

Objects to be seen on the Moon's surface range from grey plains (maria, or seas), mountain ranges, mountain-walled enclosures (such as Clavius), and mountain-ringed plains (like Copernicus) to craters and craterlets, clefts and rilles.

The 30 or so large grey areas known to the ancients as 'seas' correspond perhaps to our oceans, or perhaps to the continents – it is difficult to find a good analogy. But it is easy to understand why they were called seas, for they seem to have been covered with water aeons ago – or perhaps it was by liquid lava (molten rock). The naked-eye view of the 'Man in the Moon' is formed by these darker areas appearing against the lighter surface.

The seas are still known by their Latin names. Large ones include the following:

Mare Crisium	Sea of Crises
Mare Foecunditatis	Sea of Fecundity
Mare Nectaris	Sea of Nectar
Mare Tranquilitatis	Sea of Tranquillity
Mare Serenitatis	Sea of Serenity
Mare Frigoris	Sea of Gold
Mare Imbrium	Sea of Showers
Mare Vaporum	Sea of Vapours
Mare Nubium	Sea of Clouds
Mare Humorum	Sea of Humours
Oceanus Procellarum	Ocean of Storms

The largest of these is the Oceanus Procellarum, which occupies the eastern regions of the Moon. There are, in addition, smaller maria, like Mare Smythii (Smyth's Sea), Lacus Mortis (the Lake of Death), and Lacus Somniorum (the Lake of Dreams).

The seas are quite dissimilar in surface and extent. Small instruments give the impression that most have a level surface, but as the magnification is increased, more and more irregularities become apparent in the form of depressions, ridges, hollows and rilles. Although the maria seem quite smooth in telescopic views from the Earth, close-up photographs by the Apollo astronauts reveal small craters and occasional rilles.

Close up of the Crater Copernicus

NASA

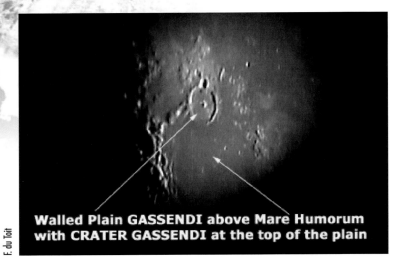

F. du Toit

**Walled Plain GASSENDI above Mare Humorum
with CRATER GASSENDI at the top of the plain**

Mare Imbrium (Sea of Showers), the largest of the 14 maria, is a great elliptical plain 1 100 km long, bounded on three sides by mountain ranges (the Carpathians, the lunar Alps), and opening on the eastern side into the Oceanus Procellarum. In the middle of the Alps is a remarkable and unique feature, as yet of unknown origin – the Alpine Valley. A narrow furrow about 8 km wide and 100 km long, some of its details are visible in a reasonably sized telescope with fairly high magnification. Almost every conceivable lunar topographic feature can be found in Mare Imbrium, the crater formations existing in just about all possible forms and sizes. A particularly bright as well as rugged area is found just south of the mountain-walled plain Archimedes.

Apollo 11, NASA

The far side of the Moon, which is always turned away from us, is rough and filled with craters. By comparison, the near side, the side we always see, is relatively smooth. The large crater (impact basin) spans about 30 km and was photographed by the crew of Apollo 11 as they orbited the Moon in 1969.

Through Earth-based telescopes, some 30 000 craters are visible, ranging from 1 km to more than 100 km across. Following a tradition established in the seventeenth century, the most prominent craters are named after famous philosophers and scientists. Craters smaller than about 1 km in diameter cannot be seen from Earth, simply because of optical limitations such as the resolving power of the telescope. Photographs from lunar orbit reveal millions of smaller craters that escape the scrutiny of Earth-based observers. Virtually all craters – both large and small – are the result of bombardment by meteoritic material.

Many of the youngest craters are surrounded by light-coloured streaks called rays that were formed by the violent ejection of material during impact. Rays emanating from the crater Copernicus, just south of Mare Imbrium, are visible through a telescope.

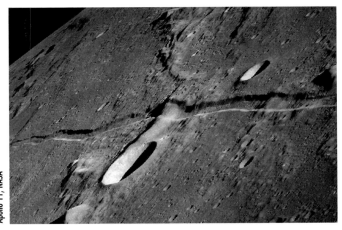

This lunar rille, the Ariadaeus rille, was photographed by the Apollo 11 crew in 1969. These long straight rilles extend for hundreds of kilometres. Three types of rilles exist on the Moon – sinuous rilles, which have wandering curves; actuate rilles, which form sweeping arcs; and straight rilles.

Apollo 11, NASA

One of the surprises of the early days of lunar exploration was the discovery that there are no maria on the far side of the Moon. The lunar far side consists entirely of heavily cratered highlands. Detailed observation by Apollo astronauts in lunar orbit demonstrated that the maria on the Moon's Earth-facing side are 2-5 km below the average elevation. In contrast, the cratered terraces on the lunar far side are typically at elevations of up to 5 km above the average lunar elevation.

Besides exploring the Moon's surface, certain special and interesting telescopic observations may be made of other phenomena involving the Moon, such as occultations and eclipses. Moving along its prescribed path against the starry background, the Moon frequently glides in front of a star or planet, hiding it from view. Because the Moon always moves in an easterly direction, the occulted object disappears behind the limb of the eastern hemisphere at immersion, and reappears in the western limb at emersion. A telescope is needed, even for bright planets occulted, since the glare of the Moon prevents the eye from noting the exact moment of disappearance or reappearance.

Last Apollo on our Moon

Lunarscape

Apollo 17

It must be remembered that an astronomical telescope reverses the image, top and bottom as well as side for side, so that a star appearing (to the naked eye) at the left of the Moon will seem to be at the right in the telescopic field.

Occultations give spectacular proof of the absence of lunar atmosphere, for the disappearance of a star is instantaneous; no gradual dimming of light occurs; suddenly it is invisible. Similarly, at the calculated time the reappearance takes place in an equally startling manner. Good observations made by amateur astronomers with proper timing are of value to certain professional astronomers, who encourage the work and use the results to improve the data of celestial mechanics.

Phases of the Moon

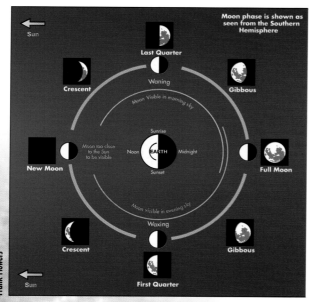

Frank Flowers

Occultation work is more interesting than it might appear and some observers become very enthusiastic about it. If you are interested in participating you should contact your local centre of the Astronomical Society of Southern Africa for more particulars.

The phase called New Moon occurs when the dark hemisphere of the Moon faces the Earth. The Moon is not visible during this phase because it is in the same part of the sky as the Sun. During the next seven days Earth-based observers see a phase called Waxing Crescent Moon in which progressively more of the illuminated hemisphere is exposed to our view. At First Quarter Moon the angle between the Sun, Moon and Earth is 90 degrees, and we see one-half of the illuminated hemisphere and one-half of the dark hemisphere. During the next week still more of the illuminated hemisphere can be seen from Earth, giving us a phase called Waxing Gibbous Moon.

When the Moon stands opposite the Sun in the sky, we see the fully illuminated hemisphere, producing the phase called Full Moon. At Full phase moonrise occurs at sunset. Over the subsequent two weeks, we see less and less of the illuminated hemisphere as the Moon continues along its orbit. This produces the phases called Waning Gibbous Moon, Last Quarter and Waning Crescent Moon.

Eclipses

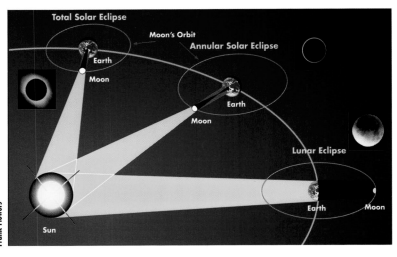

Alignments during a solar or lunar eclipse

Frank Flowers

An eclipse occurs when the Sun, Earth and Moon are aligned in space. When the Moon passes between the Sun and the Earth it casts its shadow onto the Earth's surface and we have an eclipse of the Sun, or solar eclipse. When the Earth passes between the Sun and the Moon, the Earth's shadow is cast onto the Moon and we get an eclipse of the Moon, or lunar eclipse. In any one year there is a minimum of two eclipses, of which both must be solar, and a maximum of seven, of which four are solar and three lunar, or five solar and two lunar.

Lunar eclipses

Albert Jansen

The Moon orbits the Earth roughly once a month. It takes one revolution relative to the background stars in an average period of 27.32 days. The interval between successive New Moons or Full Moons is known as the synodic period, a lunar month, or simply as one lunation. The mean period of one lunation is 29.5306 days. Therefore, one would expect that every 29.5 days, at Full Moon, conditions would be right for a lunar eclipse. As we well know, we do not get lunar eclipses at every Full Moon. Why?

The answer can be found in the fact that the Moon's orbit around the Earth is tilted slightly, by just over 5 degrees, to the orbit of the Earth round the Sun, known as the ecliptic. For an eclipse to occur, Full Moon has to be close to the point at which the Moon's orbit crosses the ecliptic – a point known as a node. In most months the Moon passes either above or below the Earth's shadow, and we see only an uneclipsed Full Moon. On occasions, however, Full Moon occurs close to a node (there are two nodes in the Moon's orbit) and the requirements for a lunar eclipse are fulfilled. The proximity of the node to the Full Moon determines what type of lunar eclipse will be seen.

Types of lunar eclipses

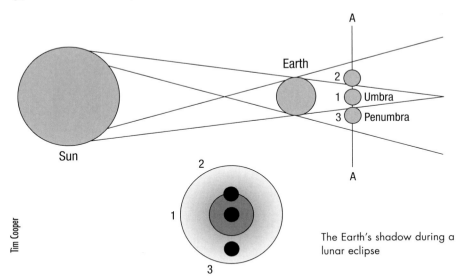

Tim Cooper

The Earth's shadow during a lunar eclipse

Lunar eclipses can be penumbral, partial or total umbral. The nature of the Earth's shadow is twofold – a dark inner shadow, called the umbra, and a lighter outer shadow, called the penumbra. If the shadow was viewed perpendicular to plane A-A it would appear (if we could view it projected onto an imaginary screen) as a dark disc surrounded by a larger lighter disc. It should be noted that the shadow is roughly circular. In fact, Aristotle noticed in the year 4BC that the shadow of the Earth on the Moon was curved, which indicated that the Earth is round.

The type of eclipse we see depends on how far from the node Full Moon occurs. If it occurs very close to the node, as in position 1, we get a total lunar eclipse; the whole of the Moon passes inside the inner dark umbra. If Full Moon occurs a little away from the node, as in position 2, the Moon's disc can still pass partially within the umbra, and we get a partial lunar eclipse. Finally, if Full Moon occurs further from the node, as in position 3, the Moon may miss the umbra, but pass within the lighter penumbra.

Magnitude and duration of a lunar eclipse

The depth by which the Moon is immersed in the Earth's shadow is called the magnitude of the eclipse. For example, a magnitude of 0.5 means that half the Moon's diameter is within the shadow at mid-eclipse. At magnitude 1.0 the entire Moon is just covered, and the eclipse is total. Thus a partial eclipse has a magnitude between 0 and 1.0. The maximum possible magnitude is about 1.6. The duration of a total lunar eclipse can be quite long. If the Moon passes through the centre of the Earth's shadow, totality can last for as long as 108 minutes. At a magnitude of 1.0 the duration of totality is a minute or less.

Circumstances of a lunar eclipse

The exact times of events in an eclipse can normally be predicted quite accurately. The times of the start and the end of eclipses are called the 'circumstances of the eclipse'. In a total lunar eclipse there are four contacts. First contact refers to the time at which partial eclipse begins; second contact refers to the start of the total eclipse; third contact is the time of the end of totality, and finally, fourth contact is the end of partial eclipse. In a partial eclipse, there is only a first and a fourth contact. Normally these contact times can be predicted accurately. However, in practice we find that first and second contact occur a minute or two earlier than predicted, while third and fourth contact occur later by the same period. This phenomenon was first noted in 1702 by a French astronomer, Pierre de la Hire. The reason is that the Earth has an atmosphere which causes its shadow to be slightly larger than predicted by simple geometry based on its diameter.

Darkness and colour of a lunar eclipse

Since the Moon is completely immersed in the Earth's shadow during a total eclipse, one would expect it to disappear from view. It hardly ever does, and remains visible during the eclipse. Once again this is because the Earth has an atmosphere.

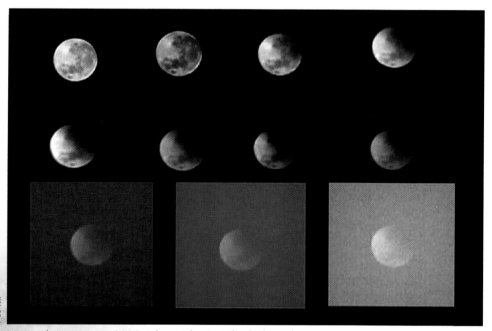

F du Toit

Lunar eclipse, January 2000, taken in the Port Elizabeth predawn

Johannes Kepler suggested in 1604 that the Earth's atmosphere acts as a lens, refracting sunlight onto the Moon. This refraction also affects the colour of the Moon during the eclipse – since blue light is scattered more than red light by particles in the atmosphere, red light is refracted onto the Moon, which appears reddish in colour. The darkness and colour vary from eclipse to eclipse, depending on conditions in the atmosphere at the time.

The phenomenon with the greatest effect on the darkness of an eclipse has been found to be dust and gas ejected into the atmosphere during volcanic eruptions. Kepler, in 1588, attributed the near disappearance of the Moon during that year's eclipse to the dust in the atmosphere. When a series of dark eclipses followed the eruption of Krakatoa in 1883, the link between eclipse darkness and volcanic eruptions was proved. These days, we can use the darkness of the eclipse as a measure of the opacity of the atmosphere caused by volcanic eruptions. If a lunar eclipse occurs fairly shortly after a major volcanic eruption one can expect the Moon to become quite dark, and possibly almost disappear from view. If an eclipse occurs several years after the last volcanic eruption, one can expect the Moon to remain bright.

Lunar eclipses

Date	Type
2005, 17 October	Total
2006, 7 September	Partial
2007, 3 March	Total
2007, 28 August	Total
2008, 21 February	Total
2009, 31 December	Partial
2010, 26 June	Partial
2010, 21 December	Total

Solar eclipses

Every 29.5306 days on average we get a New Moon, and a possible solar eclipse. But for the same reasons that we do not get a lunar eclipse every month, we do not get a solar eclipse once a month. Solar eclipses are more common than lunar eclipses, but they are visible over a smaller area of the Earth. Total eclipses are rare – the last total solar eclipses in South Africa were in 1940, when the path crossed the southern Cape, and in 2002, when the path crossed the far north of Limpopo Province.

Solar Eclipse, Hawaii

Types, duration and circumstances of a solar eclipse

A solar eclipse can be partial, annular or total. It can last as long as four hours from first to last contact, but, unlike a lunar eclipse, the duration of totality is very short, lasting at most 7 minutes and 40 seconds. During an annular eclipse, where, because the Moon is further from the Earth, its angular diameter is smaller, the duration of annularity may last as long as 12 minutes and 30 seconds.

The times of events in a solar eclipse are also expressed as circumstances. The time of start of a partial solar eclipse is first contact; the start of totality is second contact; the end of totality is third contact, and the end of partial solar eclipse is fourth contact. A partial solar eclipse has only a first and fourth contact. Because the Moon has no atmosphere and appears as a very sharp-edged object against the bright Sun, these times can be predicted very accurately, unlike those of a lunar eclipse.

Observing a solar eclipse

It must be stressed that looking at the Sun with the unprotected eyes, even during a solar eclipse, is extremely dangerous, and will result in permanent eye damage. Even during the latter stages of a partial solar eclipse or during an annular eclipse there is enough sunlight to damage the sensitive eye tissue. NEVER look at the Sun through a telescope or binoculars. The safest way to view the Sun is by projecting its image through a pinhole onto a white screen, or by using the telescope to project the image.

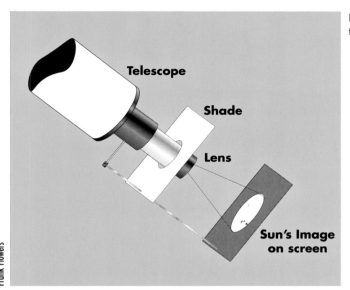

Projecting an image of the Sun

Telescope

Shade

Lens

Sun's Image on screen

Frank Flowers

At the moment of first contact you will just start to note a sharp 'bite' taken out of the Sun's disc. As the partial phases continue the 'bite' grows and the shadow will obscure any sunspots visible on the Sun's surface. If the magnitude of the eclipse is large enough to obscure more than 50 per cent of the Sun you might note that the daylight is a little dimmer than normal. After maximum partial eclipse the shadow starts to recede from the Sun, the obscured sunspots become visible again, and finally at fourth contact the last bite of shadow disappears.

If you are lucky enough to be in the track of a total solar eclipse you will see very different aspects of it. As totality nears, you will note that it has become strangely dark, and there is an almost eerie light. Birds and animals are fooled into thinking that night is approaching. As the last glint of sunlight recedes it may manifest itself in the well-known 'Diamond Ring' effect. At the instant totality begins, look for 'Baily's Beads', which are caused by sunlight filtering through valleys on the Moon's limb. At the same time, 'Shadow Bands' or fringes may be visible as bands of shadow about 10-15 cm wide and about a metre apart crossing any light surface. They are best seen by laying a white sheet on the ground.

During totality, the absence of sunlight allows us to see features which are normally invisible. This includes the Sun's inner corona – light-coloured streamers carried out from the Sun's surface, and pink-coloured prominences – huge jets of material which shoot off the surface of the Sun and loop back to crash onto the surface again.

Johannesburg Planetarium

Progress of a total solar eclipse from first contact to the moment of totality

Forthcoming solar eclipses

Date	Type	Where Visible
2005, 8 April	Annular/Total	Pacific Ocean, Central America
2005, 3 October	Annular	Atlantic Ocean, Spain, Africa
2006, 29 March	Total	Atlantic Ocean, Africa, Turkey
2006, 22 September	Annular	NE of South America, Atlantic
2008, 7 February	Annular	South Pacific, Antarctica
2008, 1 August	Total	North Canada, Arctic Ocean, China
2009, 26 January	Annular	Atlantic & Indian Oceans, Sumatra
2009, 22 July	Total	Asia, Pacific Ocean
2010, 15 January	Annular	Africa, South East Asia
2010, 11 July	Total	Pacific Ocean, South America

Eclipse information available from:
http://sunearth.gcsf.nasa.gov/eclipse/
http://umbra.nascom.nasa.gov/eclipse/

The Planets

Until Kepler's time, astronomers had assumed that heavenly objects moved in circles. Circles were considered the most perfect and harmonious of all geometric shapes. God was assumed to be in heaven along with the stars and the planets and, since God is perfect, He would only use circles to control the motions of the planets. Kepler doubted such arguments and his first major contribution to astronomy was the suggestion that non-circular curves might fit the planetary orbits.

Kepler turned from circles to ovals. For years he tried in vain to fit ovals to the orbits of planets about the Sun. Then he began working with a slightly different curve, called an ellipse. The ellipse has two foci. The longest diameter across an ellipse passes through both foci and is called the major axis. Half of the distance is called the semi-major axis, whose length is usually designated by the letter α (alpha).

To Kepler's delight, the ellipse turned out to be the curve he had been searching for. He published the discovery in 1609 and it is now known as 'Kepler's First Law': The orbit of a planet about the Sun is an ellipse with the Sun at one focus.

Kepler also realised that planets do not move at a uniform speed along their orbit. A planet moves more rapidly when it is nearest the Sun, at a point on its orbit called perihelion. It moves more slowly when it is furthest from the Sun, at a point called aphelion. After much trial and error, Kepler discovered a way to describe how fast a planet moves along its orbit. This discovery is now called the 'Law of Equal Areas' or 'Kepler's Second Law': A line joining a planet and the Sun sweeps out equal areas in equal intervals of time.

Kepler was fascinated by many harmonious relationships in the motion of the planets. His writing is filled with intriguing speculations, including musical scores intended to represent celestial music made by the planets as they travel along their orbits. One of his later discoveries stands out because of its impact on future developments. It gives a relationship between the sidereal period of a planet and the length of its semi-major axis and is now called the 'Harmonic Law' or 'Kepler's Third Law': The squares of the sidereal period of the planets are proportional to the cubes of their semi-major axes.

Kepler's Laws have a wide range of practical applications. They are obeyed not only by the planets circling the Sun, but also by artificial satellites orbiting the Earth and by two stars revolving about each other in a double star system.

To the ancient observers there were only five planets (or 'wandering stars'), moving about among what they called the 'fixed stars'. They were different from the stars in more ways than one. Not only did they wander, always staying within certain constellations (later to be called the zodiac), but they shone with a steady light that was somewhat different from the twinkling of the stars. Early astronomers knew very little about them, and it was not until Galileo in 1610 turned his newly invented telescope on the skies that they began to learn. Slowly, but very slowly, astronomers began adding to the number of the planets so that we now know nine in addition to the Earth.

Almost everything we know about the motions of the planets round the Sun and their relative distances was deduced from observations of the positions of the planets against the background of stars. This is the method by which modern knowledge of the solar system has been gained over many centuries, and anyone with sufficient patience could probably repeat the procedure.

In their physical properties (see below for basic data) the planets fall into two classes – the four inner planets and the four outer planets, with Pluto being the exception.

The diameter of a planet can be computed from its apparent angular diameter and distance from Earth. Determining its mass is more difficult, but it can be calculated if a planet has a moon (or satellite). The planet's moon obeys Kepler's Third Law, and astronomers can measure the satellite's period and semi-major axis. If a planet, Venus for example, does not have a satellite; astronomers must rely on a comet or spacecraft that passes near the planet. The planet's gravity (which is directly related to its mass) produces a deflection and, using Newtonian mechanics, astronomers can determine its mass.

Several web-sites are available that give the position of the planets in the sky.

Sun

Mercury

Venus

Earth

Mars

Jupiter

The sizes of the planets compared to the Sun's disc.
This illustration is not to scale.

Saturn

Uranus   **Neptune**

Pluto

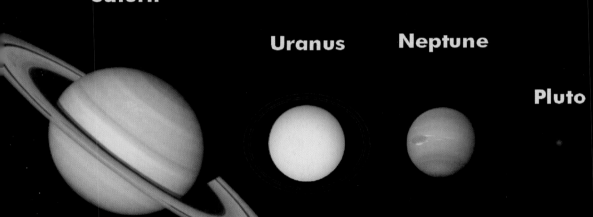

Mercury

NASA

Mercury's surface.

Basic data

Mean distance from the Sun	57 910 000 km
Mean orbital velocity	47.9 km/sec
Inclination of axis	0 degrees
Revolution period	87.969 days
Rotation period	58.81 days
Diameter	4 878 km
Mass	0.055 (Earth = 1)
Surface gravity	0.38 (Earth = 1)
Surface temperature	350° C (day)
	-180° C (night)
Escape velocity	4.3 km/sec
Albedo (percentage of Sun's radiation reflected from surface)	0.06
Brightest magnitude	-1.9

This planet, orbiting closest to the Sun, is notorious for being hard to see, and is not of much interest to the amateur observer, but it can be found more easily than you would think. Many people, in fact, have probably seen it low in the twilight and thought it was just another star.

Mercury can only be seen low in the east before sunrise, or low in the west after sunset, because its maximum elongation (the angle between the Sun and a planet as viewed from Earth) is only 27 degrees. Therefore it never rises more than two hours before sunrise or sets more than two hours after sunset.

Unfortunately, the tilt of the Earth's axis and the inclination of Mercury's orbit to the ecliptic often place the planet less than 28 degrees from the horizon at the moment of sunset or sunrise. This means that some of the elongations are 'favourable' whereas others are 'unfavourable'. Six or seven greatest elongations occur, but typically, only a couple of them will be favourable for viewing the planet.

The most practical thing to do with Mercury is to spot it. With the naked eye, look for a bright, slightly pinkish 'star'. Through binoculars or a telescope Mercury will appear as a small disc. Like the Moon, it goes through phases.

Mercury's orbital velocity is greater at perihelion (nearest the Sun), at 59 km/sec, and least at aphelion (farthest from the Sun), at 39 km/sec, varying in accordance with Kepler's Second Law. As seen from the planet's surface, the Sun rises in the east and sets in the west, just as it does on Earth. When Mercury is near perihelion, however, its rapid motion along its orbit outpaces the leisurely rotation about its axis. The usual east-west movement of the Sun through the sky is interrupted. The Sun actually stops and moves backwards (from west to east) for a few Earth days (during which temperatures are so high that lead would melt?). If you were standing on Mercury watching a sunset near the time of perihelion passage, the Sun would simply not set. It would dip below the western horizon and then come back up, only to set a second time a day or two later.

Because of Mercury's small mass, there is virtually no atmosphere. Its surface is barren, crater-scarred and mountainous, similar to that of the Moon. The daytime temperature is about 350° C, and the night is cold and daunting – about -180° C.

So far Mercury has been visited by only one spacecraft – Mariner 10 – which flew past the planet in 1974 and 1975. It would be unsafe for the Hubble Space Telescope to be used to photograph its surface as, because it is so close to the Sun, heat and radiation would damage its optical components.

Venus

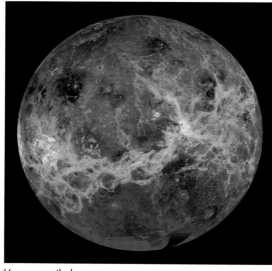

NASA

Venus unveiled

Basic data

Mean distance from the Sun	108 200 000 km
Mean orbital velocity	35.03 km/sec
Inclination of axis	178 degrees
Revolution period	224.701 days
Rotation period	-243.69 days (-Retrograde motion)
Diameter	12 100 km
Mass	0.815 (Earth = 1)
Surface gravity	0.903 (Earth = 1)
Escape velocity	10.4 km/sec
Mean surface temperature	480° C 33° C (cloud-tops)
Albedo	0.76
Brightest magnitude	-4.7

Venus, named by the ancients in honour of the Goddess of Beauty, is far brighter than any other star or planet, and can even cast a strong shadow. It is easily recognisable by its great brilliance and its characteristic pearly white colour, which is caused by the reflection of sunlight from the top of an opaque cloudy atmosphere. Very keen-sighted people claim that they can see its phases with the naked eye during the crescent stage, and binoculars show them easily.

Not only is Venus far more conspicuous than Mercury, it is also visible for longer periods. Though it is often out of sight on the other side of the Sun, these periods of invisibility are not nearly as frequent as those of Mercury. Venus can be seen for as long as four hours at a time. It slides down the celestial vault, trailing the Sun in the evening ('evening star'), and weeks later climbs into the sky before the Sun in the early hours of the morning ('morning star').

Telescopically, Venus is a great disappointment. Its dense, cloud-laden atmosphere means that from Earth the disc generally appears blank. Its atmosphere, instead of nitrogen and oxygen like that of the Earth, is basically carbon dioxide. Binoculars with a magnification of about 10 diameters and upward will reveal phases like those of Mercury and the Moon. The thin crescent is quite a striking sight, because at that time Venus is most brilliant.

Radar measurements have shown that the rotation period is 243.16 days – longer than Venus's 'year'. The planet rotates from east to west, in the opposite direction to the rotation of Earth. If it were possible to see the Sun from the surface of Venus, it would rise in the west and set in the east 118 Earth-days later. Hence a calendar on Venus would indeed be very strange. The reason for this retrograde motion is as yet unknown.

Venus has no moon.

The first spacecraft ever to visit Venus was Mariner II in 1962. Since then more than 20 have peered at the second planet from the Sun.

Earth

Earth

Basic data

Mean distance from the Sun	149 597 870 km (=1 AU)
Maximum distance from the Sun	152 100 000 km
Minimum distance from the Sun	147 100 000 km
Mean orbital velocity	29.8 km/sec
Inclination of axis	23.44 degrees
Revolution period	365.256 days
Rotation velocity of surface	1660 km/h
Inclination of equator to ecliptic	23 degrees 27 min
Diameter at equator	12 756 km
Circumference at equator	40 076 km
Circumference at pole	40 008 km
Mass	5.98×10^{24}
Escape velocity	11.2 km/sec
Surface temperature	-70° C to +55° C
Earth at perihelion (furthest from Sun)	January
Earth at aphelion (nearest Sun)	July
Autumn equinox	20/21 March
Solstice (winter)	21 June
Spring equinox	23 September
Solstice (summer)	21/22 December

Autumnal equinox marks the moment when the autumn/winter begins in the southern hemisphere as the Sun moves northward across the celestial equator (March 20/21). The spring (or vernal) equinox marks the beginning of spring/summer in the southern hemisphere as the Sun moves southwards across the celestial equator. It occurs on 23 September each year except in leap years when the extra day added in February moves it to 22 September.

Between the autumnal and vernal equinoxes there are two other significant locations along the ecliptic. Winter solstice (21 June) is the time of the year when the Sun appears at its most northerly position – over the tropic of Cancer, at latitude 23.5 degrees north. Six months later at the time of the summer solstice (21/22 December) the Sun is at its highest position and is overhead at a latitude of 23 degrees 30 minutes

NASA

Earth photographed from Apollo 17

south (the Tropic of Capricorn). Because of the procession of the equinoxes and the westward motion of the points of reference in the sky, the Sun is no longer in these constellations when overhead. The term, however, is still used.

Mars

HST

Mars

Basic data

Mean distance from the Sun	227 940 000 km
Mean orbital velocity	24.13 km/sec
Inclination of axis	23.98 degrees
Revolution period	1.88 years (686.980 days)
Rotation period	24.623 hours
Diameter	6 794 km
Mass	0.107 (Earth = 1)
Surface gravity	0.380 (Earth = 1)
Escape velocity	5.03 km/sec
Surface temperature	-140° C to -20° C
Albedo	0.16
Brightest magnitude	-2.8

Mars is an excellent object for small telescopes. It shows its famous reddish colour to the naked eye and even in a small telescope, under the right 'seeing' conditions, it offers certain surface details. Its apparent brightness varies greatly according to its position relative to Earth.

The two areas at the south and north poles, which are covered with hoar frost, can be seen through a telescope as white spots. The axis of rotation of Mars is tilted, just as the axis of the Earth is tilted, so that the planet has seasons during which the size of these frozen polar caps change. Dark areas on the surface are permanent, and when Mars is fairly close to the Earth, a small telescope will show some of these areas.

Mars has a very thin atmosphere and the average pressure on the surface is only about 7 millibars, but it can vary and get thick enough to support the very strong winds and vast dust storms that can at times engulf the entire planet.

The first spacecraft to visit Mars was Mariner 4 in 1965. Several others followed, including Mars 2, the first spacecraft to land on Mars, and the two Viking Landers in 1976. After a 20-year hiatus, Mars Pathfinder landed successfully in 1997. In July 2004 the Mars Expedition Rovers, 'Spirit' and 'Opportunity', landed on the planet, sending back geologic data and many photos. These reveal that Mars is a rocky, cold and sterile world with a hazy pink sky. It is a wasteland, a formerly volatile planet where volcanoes once raged, meteors ploughed deep into its surface creating craters, and flash floods rushed over the land.

Except for the Earth, Mars has the most varied and interesting terrain of any of the terrestrial planets, some of it quite spectacular.

Mars's two moons are named Phobos ('Fear') and Deimos ('Panic'), after the mythological companions of the Roman war god, Mars, or, as some authorities say, the names of the horses that drew the chariot of Mars. The moons are not large and are not made round by gravity like our Moon. The moons of Mars are orbiting chunks of rock. Deimos, which is further from Mars than Phobos, orbits Mars at a distance of about 20 000 km, with a sidereal period of 30 hours 18 minutes. Phobos, nearer and larger than Deimos, orbits Mars in only 7 hours 39 minutes, moving over the Martian surface at an average distance of only 6 000 km. Both moons were discovered by an American astronomer, Asaph Hall, in 1877.

Large telescopes are required to view the two moons as they do not became brighter than eleventh and twelfth magnitude objects.

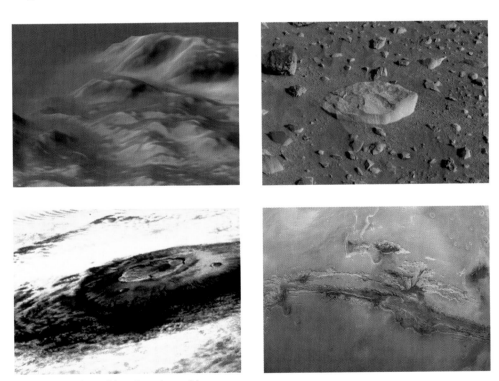

Mars ... a rocky, cold and sterile world

Jupiter

Cassini craft

Jupiter

Basic data

Mean distance from the Sun	778 330 000 km
Mean orbital velocity	13.06 km/sec
Inclination of axis	3 degrees
Rotation period	9.925 hours
Revolution period	11.9 years
Equatorial diameter	142 984 km
Mass	317.89 (Earth = 1)
Surface gravity	2.5 (Earth = 1)
Escape velocity	60 km/sec
Mean temperature (at cloud tops)	-150° C
Albedo	0.43
Brightest magnitude	-2.6

Jupiter, which orbits beyond Mars, is the largest planet in our solar system and, despite its great distance from Earth, usually appears much brighter than any other planet, with the exception of Venus.

Binoculars or a small telescope with 10-18 magnification will show a tiny disc, but a larger instrument is required to enjoy Jupiter. Through a 114 mm reflector or 90 mm refractor telescope, Jupiter resolves into a series of red, yellow, and brown shadings, as well as a wealth of telescopic detail.

Exceptionally good 'seeing' conditions are not required to get a clear view of Jupiter's markings. The cloud bands for which the planet is famous will appear in a low-powered telescope as parallel bands stretching across the disc. Another well-known feature that the amateur may be able to observe is the 'Great Red Spot', discovered in 1664 by Robert Hooke. The Spot is visible in Jupiter's southern hemisphere, where it extends over 28 000-40 000 km in longitude and 13 000 km in latitude. Although it has been shrinking over the years (and sometimes fades) it could hold three Earths like peas in a pod.

Jupiter appears to have a solid core surrounded by layers of liquid hydrogen, above which comes the 'atmosphere'. The atmosphere is believed to be made up of hydrogen together with helium and some other unpleasant hydrogen compounds such as ammonia and methane.

Like Saturn, Jupiter has rings, but these are much smaller and fainter. The rings were discovered by the Voyager 1 spacecraft, but are not visible through Earth-based telescopes. It is Jupiter's moons that contain the greatest telescopic treasure for the amateur. The four that are visible run a race with one another around the planet and change their respective positions from hour to hour and night to night. They are not hard to identify with the aid of an astronomical almanac, and they can be followed for hours as they speed in front of Jupiter, throwing their shadows on the planet, vanishing behind its giant disc, or plunging suddenly into its deep long shadow.

Io, which is covered with volcanoes, of which about ten are now erupting, is the most active body in the solar system; Europa is apparently wrapped in a deep layer of ice, underneath which may lie a huge ocean; Ganymede, which may contain water and ice surrounding a core of rock, shows craters and strange grooves; and Callisto, which is densely packed with craters, includes a huge bulls-eye impact crater called Valhalla (300 km across).

These four Galilean moons (thus named because they were studied as long ago as 1610 by Galileo, the first great astronomer) were merely points of light in the sky until NASA's two Voyager spacecraft flew nearby in 1979 and 1980, radioing back close-up pictures of their surfaces. In 1998 the spacecraft Galileo was launched from the cargo bay of Space Shuttle Atlantis to get a good glimpse of Jupiter. Then, in 1994, the spacecraft made the only direct observation of the comet Shoemaker-Levi impacting Jupiter. After a 14-year odyssey that included two years of orbiting Jupiter and intensive investigation of the planet as well as unveiling some of the mysteries of the moons, the Galileo spacecraft disintegrated in the planet's dense atmosphere. The spacecraft was purposely put on a collision course with Jupiter because its fuel was nearly depleted and without it the spacecraft could no longer be controlled.

NASA

In July 1994, Comet Shoemaker-Levy 9 collided with Jupiter, with spectacular results. The effects were clearly visible even through amateur telescopes! The debris from the collision was visible for nearly a year afterwards with HST.

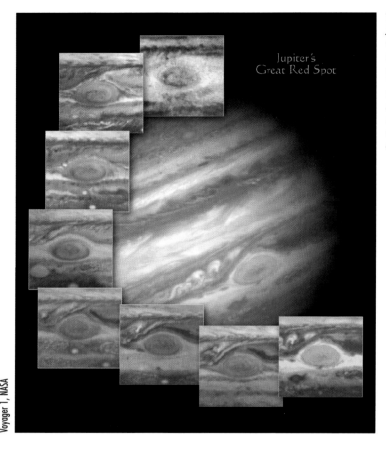

Jupiter's
Great Red Spot

This photograph of Jupiter was taken by the Voyager 1 spacecraft as it passed the planet in 1979. Clearly visible is the Great Red Spot, a giant hurricane-like storm system that rotates with the clouds on Jupiter.

Voyager 1, NASA

The planes of the orbits of the Galilean moons correspond, as does that of Jupiter itself, with the plane of the ecliptic, and therefore with our line of sight. Consequently we see Jupiter and four moons arranged in a fairly straight line, and the moons can be seen to transit Jupiter or be occulted by it. The Sun is so close to the satellite plane that these moons always cast a shadow-transit on Jupiter's surface, and suffer eclipses whenever they pass behind the planet.

Minimum telescope apertures of about 75 mm for shadows and 150 mm for satellite in transit are necessary.

Jupiter has 63 known satellites (moons) (as of February 2004): in addition to the four large Galilean moons, there are 34 smaller named ones, plus many more small ones discovered recently but not yet named. The moons of Jupiter are named for the other figures in the life of the chief god (mostly his numerous lovers).

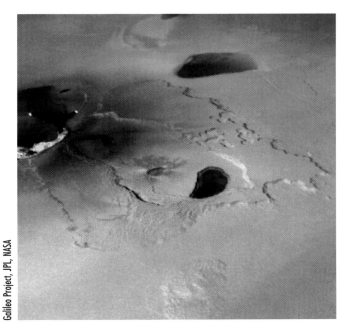

Galileo Project, JPL, NASA

A volcano on Jupiter's moon Io was photographed during an ongoing eruption. A glowing landscape of plateaux and valleys covered in sulphur and silicate rock surround the active volcano. Io is slightly larger than our own Moon and is the closest moon to Jupiter. This shows a region about 250 km across. Io's volcanoes are so active that they are causing molten rock to explode through the surface. Indeed, the volcanoes are so active they are effectively turning the whole moon inside out.

Jupiter's named moons, in order of discovery

Io	Europa	Ganymede
Callisto	Amalthea	Himalia
Elara	Pasiphae	Sinope
Lysithea	Carme	Ananke
Leda	Thebe	Adrastea
Metis	Callirrhoe	Themisto
Megaclite	Taygete	Chaldene
Harpalyke	Kalyke	Iocaste
Erinome	Isonoe	Praxidike
Autonoe	Thyone	Hermippe
Aitne	Eurydome	Euanthe
Euporie	Orthosie	Sponde
Kale	Pasithee	

Jupiter is very gradually slowing down because of the tidal forces produced by the Galilean moons. The same tidal drag is also changing the orbits of the moons, forcing them slowly farther away from the planet.

Saturn

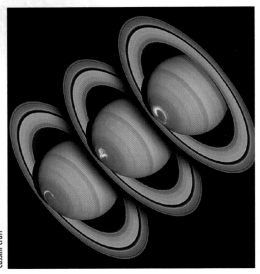

Cassini craft

Basic data

Mean distance from the Sun	1 429 400 000 km
Mean orbital velocity	9.6 km/sec
Inclination of axis	29 degrees
Revolution period	29.46 years
Rotation period	10.50 hours
Equatorial diameter	120 536 km
Mass	95.17 (Earth = 1)
Surface gravity	1.1 (Earth = 1)
Escape velocity	35.5 km/sec
Mean surface temperature	-180° C (at cloud tops)
Albedo	0.61
Brightest magnitude	-0.3

Voyager took this photograph in 1981. Saturn's famous rings are visible along with two of its moons, Rhea and Dione, which appear as faint dots on the right and lower right sections of the picture.

Saturn is not unlike Jupiter, but it is less massive, with a mean density less than that of water. Its chief glory is, of course, its ring system, which looks like a disc of material, thin and flat, as if it were cut out of cardboard, surrounding the equator of the planet. Two prominent rings (A and B) and a faint one (C) can be seen from Earth.

As Saturn moves round the Sun we see the rings at varying angles – apparently from above, below, or edge-on, according to the planet's position in relation to the Earth. The rings reflect so much light that when they present their broadest aspect to the Earth, Saturn in the sky seems three times brighter then when the rings are edge-on. And when they are edge-on (which occurs every 15 years) they are invisible in small telescopes and almost invisible even in the very largest.

At other times a medium-sized instrument will show the division of the rings quite clearly. The first one to resolve itself is the dark Cassini division, which divides the system in two and is not hard to detect with a small telescope. On the outer of the two rings thus formed, one may be able to pick up the faint, grey Encke division which can be elusive and is not always visible.

The origin of the rings of Saturn (and the other gas giants) is obscure. It is thought that the rings may have been formed from the larger moons that were shattered by the impact of comets and meteroids. The composition is not known for certain, but the rings do show a significant amount of water.

Saturn's rings

Saturn has a large family of moons – at least 30. Nine of these were known before the Voyager and Pioneer spacecraft encountered Saturn. The largest moon, Titan, is approximately 5 150 km in diameter, and the only moon in our solar system which contains a substantial atmosphere composed mainly of nitrogen, suggesting that Titan might be an 'Earth' in deep freeze. But Titan is so far away that sunlight has to travel much further to reach it than to reach Jupiter's moons, and must then travel further back to Earth. As a result Titan never becomes brighter than about eighth magnitude; it can, however, be seen with even a small telescope as a small, star-like point of light.

Saturn's moons are unfortunately not easy to observe with small instruments, and it would be a lucky observer who could pick up any other than Titan in a three-inch telescope. Equipped with a four-inch telescope, however, one might be more fortunate. Taking extreme care to distinguish them from the stars, one should be able to identify Titan, Iapetus, Rhea, Tethys, and Dione.

Phoebe

Rhea

Tethys

101

Names of Saturn's moons in order of discovery:

Mimas	Enceladus	Tethys
Dione	Rhea	Titan
Hyperion	Iapetus	Phoebe
Janus	Epimetheus	Helene
Telesto	Calypso	Atlas
Prometheus	Pandora	Pan
Ymir	Paaliaq	Tarvos
Ijiraq	Suttungr	Kiviug
Mundilfari	Albiorix	Skathi
Erriapo	Siarnag	Thrymr

One further moon, discovered in 2003, has yet to be named.

Saturn was first visited by the Pioneer 11 spacecraft in 1979, and later by Voyagers 1 and 2. They were followed in 2004 by a mission named Cassini-Huygens in honour of the Italian Jean-Dominique Cassini, who, in 1675, discovered what is now known as the 'Cassini Division', the narrow gap separating Saturn's rings, and Dutch scientist, Christiaan Huygens, who discovered Saturn's rings and its largest moon, Titan, in 1655. The Cassini-Huygens space probe is an international collaboration between NASA, the European Space Agency (ESA) and Agenzia Spaziale Italiana (ASI). The Cassini Orbiter was built and managed by NASA's Jet Propulsion Laboratory, and the Huygens probe was built by ESA, while ASI provided Cassini's high-gain communication antenna. More than 250 scientists worldwide will study the data collected.

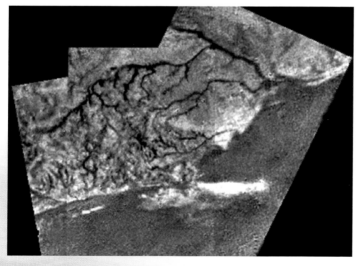

Titan's surface

The Cassini-Huygens mission was launched on 15 October 1997 and arrived at Saturn in July 2004, after a seven-year journey. The spacecraft slipped through a gap in the rings and sent back many photographs showing that the rings consist of millions of icy particles in a range of sizes. In December 2004, towards the end of Cassini's third orbit around Saturn, the Huygens probe was ejected on a 22-day cruise to Titan, about 1.2 billion kilometres from Earth, reaching the moon on 14 January 2005. Although the Huygens probe no longer operates, the images it returned will be studied for decades to come, while the Cassini mother ship will continue to orbit Saturn and return images for many years.

Cassini-Huygens is the largest interplanetary spacecraft ever built, weighing in at 5.6 tonnes.

Cassini-Huygens spacecraft

Titan is one of the most mysterious objects in our Solar System, and the second largest, with a thick, methane-rich, nitrogen atmosphere that experts believe may resemble that of our young Earth.

Uranus

Uranus

Basic data

Mean distance from the Sun	2 870 990 000 km
Mean orbital velocity	6.8 km/sec
Inclination of axis	98 degrees
Revolution period	84.01 years
Rotation period	-10.50 hours (-Retrograde motion)
Equatorial diameter	51 118 km
Mass	14.5 (Earth =1)
Surface gravity	0.8 (Earth = 1)
Escape velocity	21 km/sec
Mean surface temperature (at cloud tops)	-216° C
Albedo	0.35
Brightest magnitude	+5.6

Uranus was discovered by William Herschel in 1781, more or less by accident. While making a systematic survey of the stars, he noticed that one refused to remain in its place. After its discovery, Herschel first wished to give it the name Georgium Sidus, after King George III – thankfully the name did not find favour.

A 10 to 12-inch telescope magnifying 200-300 times is needed before Uranus shows any appreciable disc. Its blue-green colour (as is the case with Neptune) is the result of absorption of red light by methane in the upper atmosphere. Hence, for most amateur observers the planet is out of reach, but, with patience, one could try to locate it and map out its path among the stars.

Uranus has a gaseous surface, but its most curious feature is the tilt of its axis, which is 98 degrees. The Earth's axis of rotation is inclined by 23.5 degrees to its orbit, and this is why we have seasons. Tilted at 98 degrees – more than a right angle – means that as seen from the Earth, the pole of the planet would lie in the centre of the disc, with the equator extending all the way around it. Why this strange tilt? Astronomers confess that they do not know. Actually there is an ongoing battle over which of Uranus's poles is its north pole. Either its axial inclination is a bit over 90 degrees and its rotation is direct, or it is a bit less than 90 degrees and the rotation is retrograde.

Like the other gas planets, Uranus has rings. Like Jupiter's, they are very dark, but like Saturn's they are composed of fairly large particles, ranging up to 10 metres in diameter, in addition to fine dust. There are 11 known rings, all very faint; the brightest is known as the Epsilon ring. The Uranus rings were the first to be discovered after Saturn's. This was of considerable importance since we now know that rings are a common feature of the outer planets, not a peculiarity of Saturn alone.

Uranus has so far been visited by only one spacecraft – Voyager II, in January 1986.

Uranus has 21 known moons (with names drawn mostly from Shakespeare's plays).

Names of moons in order of discovery

Titania	Oberon	Miranda
Ariel	Umbriel	Puck
Cordelia	Ophelia	Bianca
Cressida	Desdemona	Juliet
Portia	Rosalind	Belinda
Caliban	Sycorax	Stephano
Prospero	Setebos	Trinculo

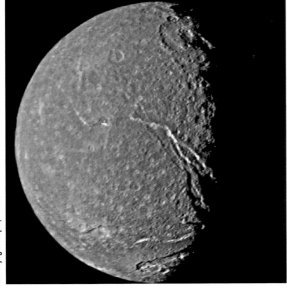

The Voyager Project, NASA.

Titania's Trenches. Voyager 2 visited Uranus in 1986 and took this picture of Titania, Uranus's largest moon. The icy and rocky moon is covered with impact craters and fault valleys visible as trench-like features. The large impact crater near the top has been given the name Gertrude and is about 150 km in diameter.

Neptune

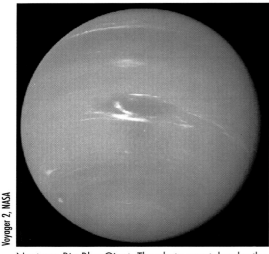

Voyager 2, NASA

Basic data

Mean distance from the Sun	4 504 000 000 km
Mean orbital velocity	5.43 km/sec
Inclination of axis	29 degrees
Revolution period	164.79 years
Rotation period	17.24 hours
Equatorial diameter	49 553 km
Mass	17.2 (Earth = 1)
Surface gravity	1.1 (Earth = 1)
Escape velocity	23.6 km/sec
Mean surface temperature (at cloud tops)	-220° C
Albedo	0.35
Brightest magnitude	+7.7

Neptune: Big Blue Giant. The photo was taken by the Voyager 2 spacecraft in 1989, the only spacecraft ever to visit Neptune.

Neptune was discovered in 1846 by Adams and Leverrier as a result of independent mathematical predictions based on irregularities in the motions of Uranus. It is not a naked eye object and can only be located with the aid of a telescope as its magnitude is about eight. It will appear as a tiny greenish/bluish disc through a telescope of about 203 mm aperture and just a dot in smaller telescopes. Its colour comes from methane in its atmosphere.

However, it offers one consolation to the amateur observer – once located it can easily be followed for the rest of the season because it spends nearly 14 years in each section of the zodiac.

Almost everything we know about Neptune comes from the visit of the Voyager II spacecraft in August 1989. At that time, its most prominent feature was the Great Dark Spot in its southern hemisphere. It was about half the size of Jupiter's Great Red Spot (about the same diameter as Earth). However, the Hubble Space Telescope's observations of Neptune in 1994 show that the Great Dark Spot has disappeared. It has either simply dissipated or is currently being masked by other aspects of the atmosphere. Voyager 2 also confirmed the existence of a complete ring system around Neptune.

Neptune has 13 known moons. Four discovered in 2002 and one in 2003 have yet to be named.

Names of moons in order of discovery

Triton	Nereid	Naiad
Thalassa	Despina	Galatea
Larissa	Proteus	

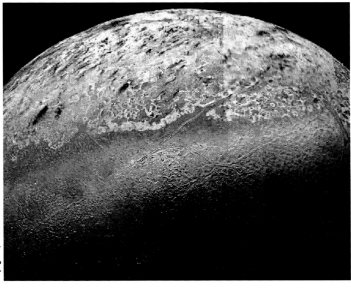

Triton: Neptune's largest moon. A fascinating terrain, with a thin atmosphere and evidence of ice volcanoes.

Voyager 2, NASA

Pluto

Pluto, Charon

Basic data

Mean distance from the Sun	5 913 520 000 km
Mean orbital velocity	4.7 km/sec
Inclination of axis	118 degrees
Revolution period	248.54 years
Rotation period	-6.387 days (-Retrograde motion)
Equatorial diameter	2 300 km
Mass	0.0025 (Earth = 1)
Surface gravity	0.08 (Earth = 1)
Escape velocity	1.3 km/sec
Mean surface temperature	about -220° C
Albedo	about 0.4
Brightest magnitude	+14

Pluto is far too small to be seen without the aid of a telescope with an aperture of at least 250 mm and is, needless to say, beyond the range of small telescopes. It was discovered in 1930 by Clyde Tombaugh. At that time it was in the constellation of Gemini.

Pluto's orbit is highly eccentric. At times it is closer to the Sun than Neptune (as it was from January 1979 to February 1999). Pluto rotates in the opposite direction from most of the other planets, and, like that of Uranus, the plane of its equator is almost at right angles to the plane of its orbit.

The surface temperature of Pluto varies between -235 and -210° C. Little is known about its atmosphere, which probably consists primarily of nitrogen with some monoxide and methane. It is extremely tenuous, the surface pressure being only a few microbars, and may exist as a gas only when Pluto is near its perihelion. For the majority of Pluto's long year, the atmospheric gases are frozen into ice.

Pluto has as company in its distant orbit a single satellite named Charon, in honour of the sinister boatman who used to ferry departed souls across the river Styx on their way to the Underworld. Discovered by James W Christ, Charon races through its orbit round Pluto in just 6.3 days, which is the same as the axial rotation of Pluto, so that the two are 'locked', and an observer on Pluto would see Charon hanging motionless in the sky.

Sedna

While astronomers were working with the 1.2 metre field telescope on Palomar Mountain in southern California during 2003 they discovered the planet now named Sedna. Since then it has also been photographed by the Hubble Space telescope. A frozen mini-world, with temperatures below absolute zero, it has a diameter of about 1600 km and is about three times as far from the sun as Pluto. To date Sedna is the most distant object to be discovered within our solar system.

There are some who think that Pluto, and now Sedna, would be better classified as large asteroids rather than planets. Some consider these two bodies to be the largest of the Kuiper Belt objects (also known as Trans-Neptunian Objects). There is considerable merit to the latter theory, but historically Pluto has been classified as a planet and it is very likely to remain so. Does the same hold for Sedna?

Proplyds

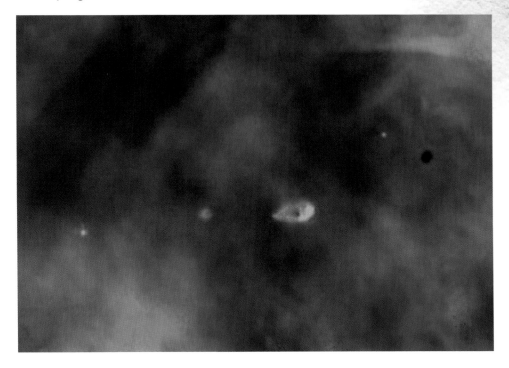

In 1994 the Hubble Space Telescope provided this image of the so-called 'proplyds'. These proplyds reveal what seem to be discs of dust and gas surrounding newly formed stars, fuzzy blobs which appear to be infant solar systems in the process of formation. The area photographed is a close-up of the Orion Nebula. Of the five stars in this field, four appear to have associated proplyds.

The Asteroids

In addition to the nine planets and their moons the solar system contains a large number of lesser planets or asteroids. In a great broad zone between the orbits of Mars and Jupiter – the region known as the 'asteroid belt' – and perpetually circling the Sun in ellipses, are thousands of small bodies, each a world in itself. These are the asteroids or minor planets. They are forever beyond the range of the naked eye, yet it is not difficult to locate a few of them with low telescopic power – even binoculars – and to follow them, night after night, during an entire apparition.

The immense zone containing all these minor planets is a great ring-shaped belt. While the orbits of most of the asteroids are confined to this area, an occasional one is found beyond the outermost boundary or within the innermost limit.

The first four asteroids that have been discovered are sometimes called the Big Four, as they are the largest and brightest. Observers with even small telescopes can locate any of them when they are in the night sky.

Ceres, the first and largest of the Big Four, was discovered by Piazzi in 1801. Its diameter is a scant 780 km; its average magnitude is about 7.4 and it orbits the Sun once every 4.6 years. This is in remarkable agreement with the distance 'predicted' by Bode's Law for a missing planet.*

In 1802, the second largest asteroid, Pallas, was found by Olbers. It has a diameter of 450 km and a magnitude of 8.0. The metal palladium was named after Pallas. The third to be discovered was Juno, found by Harding in 1804. It has a diameter of 250 km, making it the smallest of the Big Four. Its magnitude is about 8.7. In 1807 Olbers found Vesta, the brightest and most easily observed, with a diameter of about 380 km and an average magnitude of about 6.5.

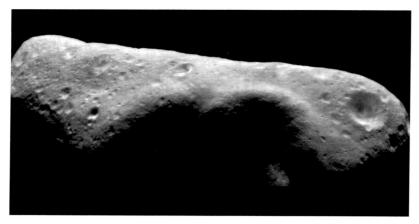

Eros

It is possible to begin observing any one of the Big Four asteroids at the start of an apparition and to follow it to the end of visibility. Start with the ephemeris of the asteroid. This is a table of the exact positions of the object at regular intervals – positions given in coordinates of right ascension and declination. Most astronomical computer programmes will list and indicate the positions of the Big Four on their star chart.

*The German astronomer, Johann Bode, who proposed the name of Uranus for the new planet which William Herschel had discovered in 1781, is far more famous for formulating a simple rule describing the distances of the planets from the Sun. Although it is not a 'law' in the true sense, merely a coincidence, it did lead directly to the discovery of a large number of previously unknown objects that orbit the Sun.

As it crossed the asteroid belt in October 1991, the spacecraft Galileo took this image of Gaspra, the first ever close-up photograph of an asteroid.

Telescopically, asteroids are not spectacular. They cannot be distinguished by appearance from stars of the same magnitude, except that they possibly sparkle less. Some – like Vesta, which appears slightly orange or pinkish – do have characteristic colours. However, their daily motion against the stellar background is a positive indication of their character. A telescope user invariably gets much practice and fun – perhaps even a thrill – from following the movements of these tiny worlds, most of which are less than 1 km across.

The final result is a series of numbers that corresponds remarkably well to the distance (in Astronomical Units) of the planets from the Sun.

Astronomers regarded Bode's Law merely as a useful trick for remembering the planetary distances from the Sun until Herschel's unexpected discovery of Uranus very near the orbit predicted by Bode for a planet beyond Saturn. Astronomers now looked with new interest at the 'missing planet' between the orbits of Mars and Jupiter.

There are roughly 100 000 asteroids bright enough to appear on Earth-based photographs. Their combined matter would produce an object barely 1 500 km in diameter – considerably smaller than our Moon. There is little support today for the hypothesis of a shattered planet. It seems more reasonable to suppose that the asteroids are debris left over from the formation of the solar system out of the solar nebula.

Bode's rule for remembering the distance of the planets from the Sun:
- Write down the sequence of numbers 0, 3, 6, 12, 24, 48, 96, …(note that each number after 3 is simply twice the preceding number)
- Add 4 to each number in sequence
- Divide each of the resulting numbers by 10.

Recommended further reading

Burnham, Robert. 1978. *Burnham's Celestial handbook: An Observer's Guide to the Universe Beyond the Solar System*. Dover.

Dickinson, Terence & Jack Newton. 1998. *Splendours of the Universe: A Practical Guide to Photographing the Night Sky*. Firefly Books.

Edberg, Stephen J & David H Levy. 1994. *Observing Comets, Asteroids, Meteors and the Zodiacal Light*. Cambridge: Cambridge practical Astronomy Handbooks.

Meeus, Jean. 1995. *Astronomical Tables of the Sun, Moon and Planets 2nd Edition*. Sky Publishing.

Moore, Patrick. 2002. *The Sky at Night*. Philip's.

Newton, Jack & Philip Teece. 1995. *The Guide to Amateur Astronomy*. Cambridge: Cambridge University Press.

O'Meara, Stephen. 2000. *The Messier Objects*. Cambridge: Cambridge University Press.

Ridpath, Ian. 2003. *Norton's Star Atlas and Reference Handbook*. Pi Press.

Rukl, Antonin. 2004. *Atlas of the Moon*. Sky Publishing.

Tirion, Wil & Patrick Moore. 1997. *The Cambridge Guide to Stars and Planets*. Cambridge: Cambridge University Press.

Tirion, Wil & Brian Skiff. 2002. *Bright Star Atlas*. Willman-Bell